Beneath the Skin, Scale, Fur, & Feather

A coloring book introduction to animal anatomy for future veterinarians

Dr. Chris Carpenter
with Jeff Thomason, PhD

Illustrations by Rebecca Brebner

Brought to you by

Vet Set Go!™ is a trademark of Vet Set Go, LLC.

Beneath the Skin, Scale, Fur, & Feather
Copyright © 2021 by Dr. Christopher Carpenter

All rights reserved. No part of this book may be used or reproduced in any manner whatsoever without written permission except in the case of brief quotations embodied in critical articles and reviews.

Printed in the United States of America.

For more information, visit our website at www.vetsetgo.com

Acknowledgments

We acknowledge the following people and organizations for their support of aspiring veterinarians.
Without their contributions, this book would not have been possible.

Brent Mayabb, DVM   JCQ   **Teresa Troutman Lind,** DVM

I am so grateful to Think Partners for never giving up and staying with me on this multi-year journey.

Animal anatomy is amazing, and we're going to show you why!

Inside this book, we will share some of the most important structures in an animal's body. We'll also share how these same structures can be found in a variety of animals that live on land, in the air, and in the water. You will see how the organs, muscles, and bones have changed to fit the needs of each animal! More importantly, we are going to explain how these different structures work and how they help animals thrive in their unique environments.

Pages 2-7
CHAPTER 1
The Skull

Pages 8-13
CHAPTER 2
The Arms

Pages 14-19
CHAPTER 3
Eating & Digesting

Pages 20-27
CHAPTER 4
The Eyes & Ears

Pages 28-33
CHAPTER 5
The Neck Bones

Pages 34-39
CHAPTER 6
The Heart

To make the most of the book, here's what we really want you to do:

Use colored pencils.
We made the coloring pages thick and durable. They will work with markers. However, we have found that the best results come with colored pencils. They shade the illustrations without covering them up. For best results, PLEASE USE COLORED PENCILS.

Use the same color for the same body part in each chapter.
Structures are numbered in each chapter to help you learn. When you pick a color for a body part, use that color for the rest of the chapter. For example, if you color the humerus of the dog blue in Chapter 2, use that same blue for the humerus of the horse, alligator, bird, bat, and dolphin too! It will let you see how that bone has adapted to serve each animal's needs.

Take the time to read and understand all that we have written for each chapter.
We don't just show the anatomy of each animal. We explain how many of the body parts work and how they have adapted in each animal. When you take the time to read through the pages, you'll understand not only where bones, organs, and other body parts are located, but also what they do inside the animal.

Are you ready?

Turn the page, and start your animal anatomy journey!

Dr. Chris Carpenter

The Skull

This is a dog's skull (you're looking down at the top of its head). Can you find the eyes? When you look at an animal, the eyes are often the first thing you notice. The eyes—and other organs like the ears, nose, tongue, and brain—are housed in the skull.

But the skull is much more than just a bony box to hold organs. It also:

- Has teeth for feeding
- Protects the brain
- Houses the special senses, like the eyes and nose, that tell the animal who might be coming to eat it or where its own food is
- Has horns or antlers for defense or fighting in some animals

Notice how each skull bone is identified by a specific number? Shade bones with the same number in the same color on every drawing in this chapter—then you'll see how bones change in size and shape in different animals.

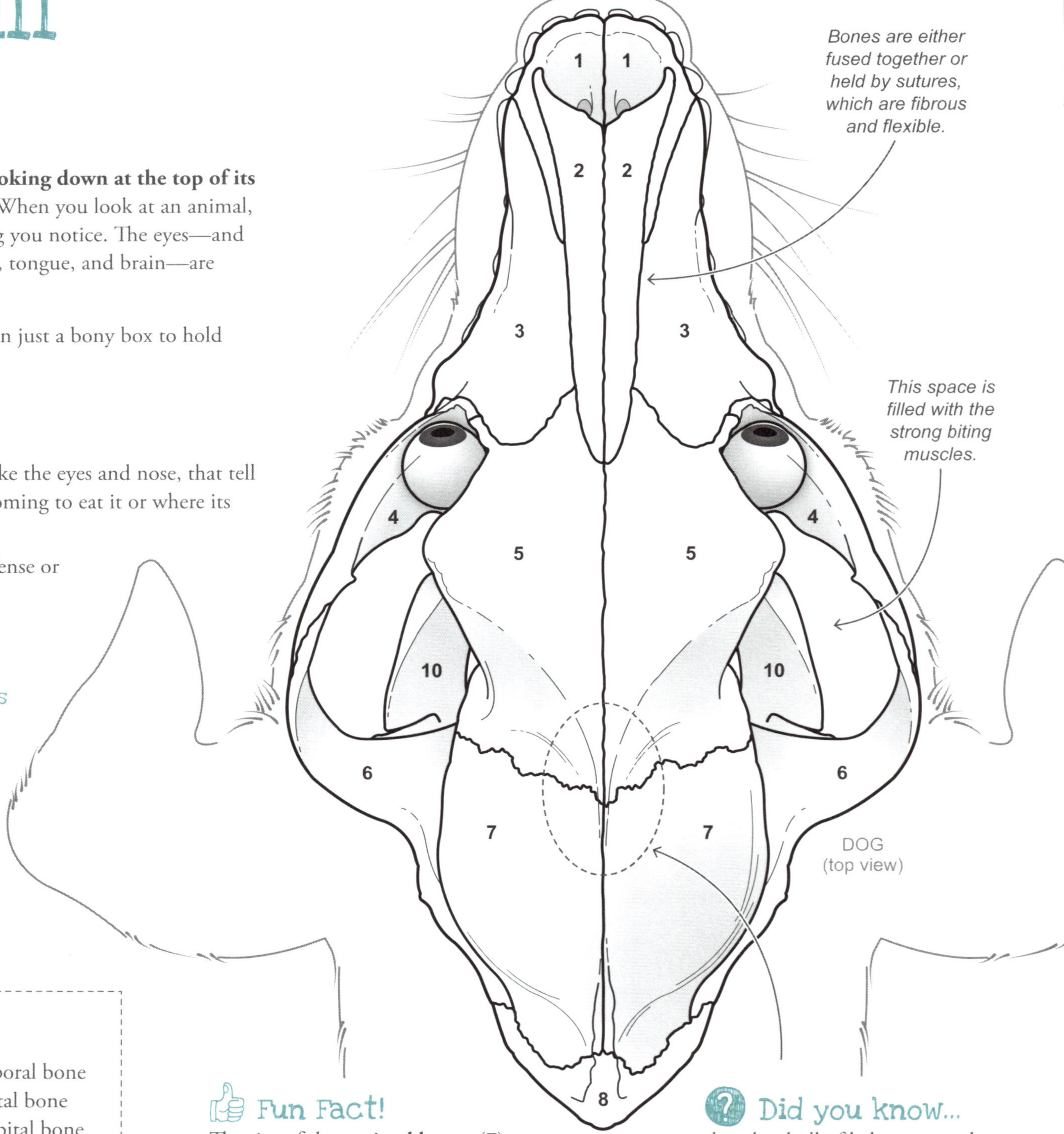

Bones are either fused together or held by sutures, which are fibrous and flexible.

This space is filled with the strong biting muscles.

DOG (top view)

Bone Names

1. Incisive bone
2. Nasal bone
3. Maxillary bone
4. Zygomatic bone
5. Frontal bone
6. Temporal bone
7. Parietal bone
8. Occipital bone
9. Lacrimal bone
10. Mandible

👍 Fun Fact!
The size of the **parietal bones** (7) and the back half of the **frontal bones** (5) show how big the dog's brain is.

❓ Did you know...
that the skull of baby mammals has a soft spot that helps it shrink slightly? This makes it easier for the mother to squeeze the baby out during birth.

The Skull

This is a dog's skull from the side. In this view, you can see that:
- Bones 1, 3, and 10 hold the teeth and, with bone 2, make the face.
- Bones 4, 5, and 9 surround the eye.
- Bones 5, 6, 7, and 8 protect the brain.

✏️ Color the bones here with the same colors you used for the bones on the opposite page. Keep the teeth white, or shade them any color you'd like!

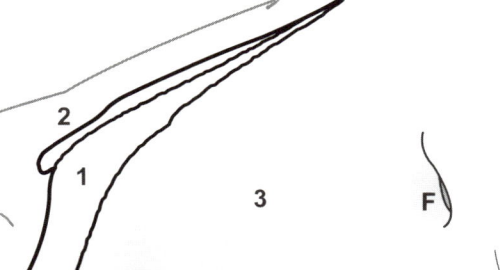

DOG (side view)

The ear hole allows sound waves to get to the hearing organ.

In this space are some hard-to-see bones with interesting names like pterygoid, ethmoid, and sphenoid. The brain sits on the last two.

👍 **Fun Fact!**
A hyena's teeth are much harder than ours—so hard that they can bite right through bones!

❓ **Did you know...**
that the skull has lots of holes? These holes are called **foramina** (a single hole is a foramen). They allow nerves, arteries, and veins to pass into and out of the skull. Find some, labeled "F," on this drawing.

❓ **Did you know...**
that skull bones are a little like building blocks and a little like jigsaw pieces? They are bigger or smaller in different animals depending on how each one is used.

The Skull

The skull of an elephant is a tall box to make it strong for attaching the heavy trunk and tusks, and for holding the large teeth. It's not as heavy as it looks because the bones have air-filled spaces in them, called **sinuses**. So do some of ours.

😕 Did you know...
that tusks (which are similar to teeth) can get cavities, just like our own teeth? So some lucky veterinarians get to do dentistry on elephants (but not on shrews)!

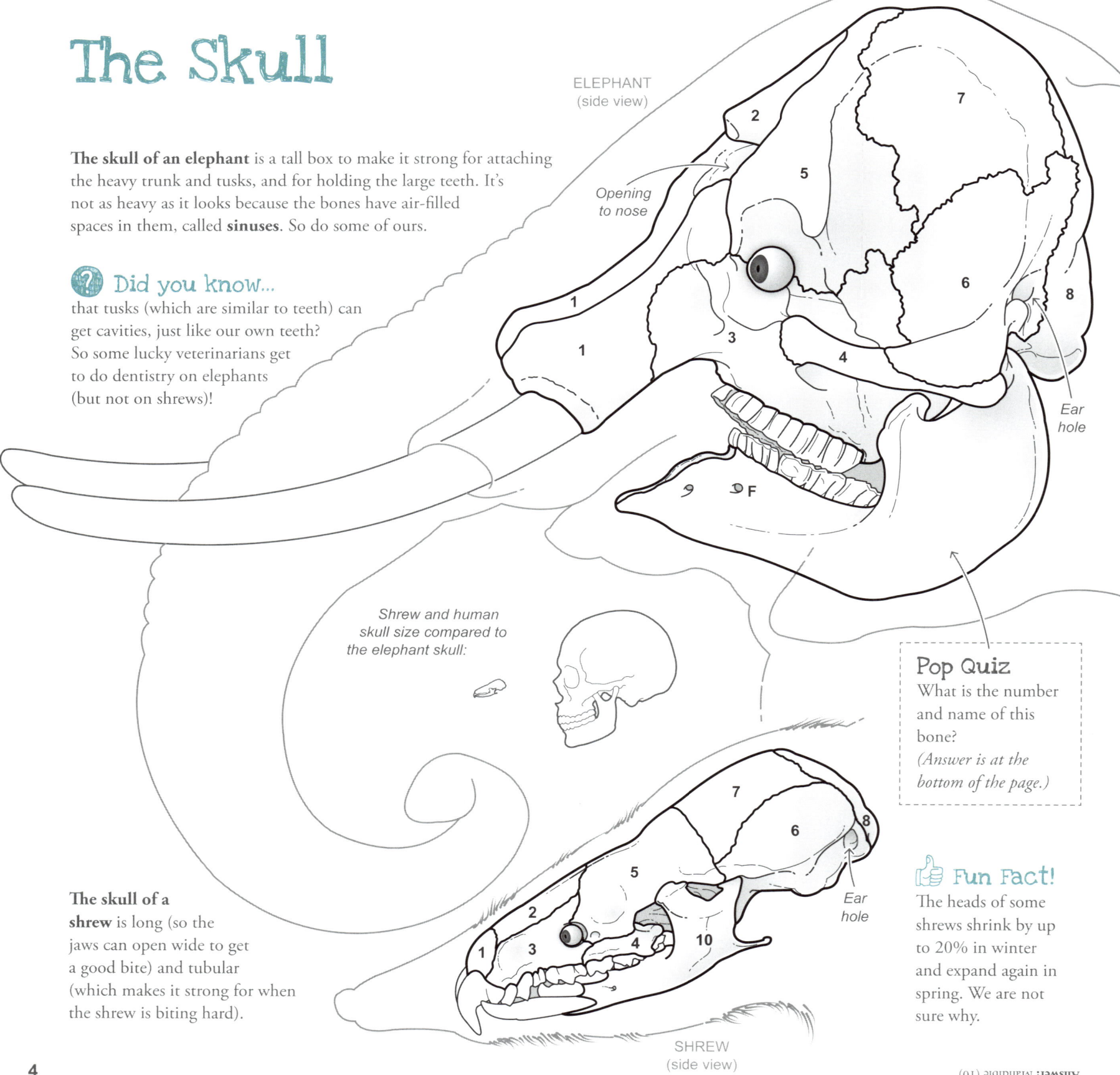

ELEPHANT (side view)

Opening to nose

Ear hole

Shrew and human skull size compared to the elephant skull:

Pop Quiz
What is the number and name of this bone?
(Answer is at the bottom of the page.)

The skull of a shrew is long (so the jaws can open wide to get a good bite) and tubular (which makes it strong for when the shrew is biting hard).

SHREW (side view)

👍 Fun Fact!
The heads of some shrews shrink by up to 20% in winter and expand again in spring. We are not sure why.

Answer: Mandible (10)

The Skull

✏️ Color the skulls the same bone colors you used for the dog so you can see how each bone changes shape in the elephant, shrew, anteater, and camel.

The skull of a giant anteater has a very long face (bones 2, 3, and 10) but no teeth! The face is a tube to house its even longer ant-licking tongue.

👍 **Fun Fact!**
Giant anteaters eat 30,000 ants or termites a day! Have you ever eaten even 1 ant?

The skull of a camel makes a long face (bones 3 and 10) to house its large teeth for biting and for grinding up its food.

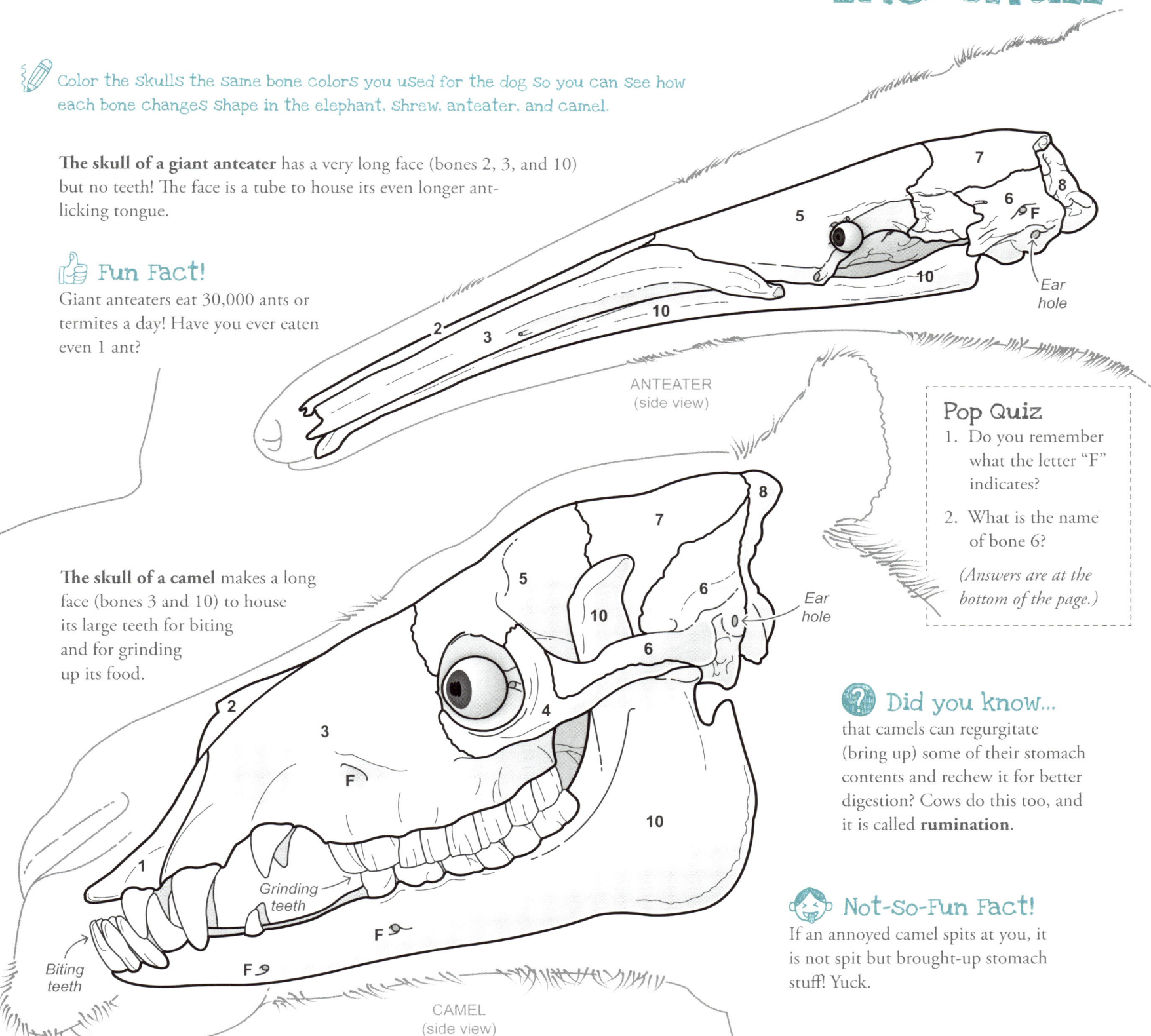

ANTEATER (side view)

CAMEL (side view)

Pop Quiz
1. Do you remember what the letter "F" indicates?
2. What is the name of bone 6?

(Answers are at the bottom of the page.)

❓ **Did you know...**
that camels can regurgitate (bring up) some of their stomach contents and rechew it for better digestion? Cows do this too, and it is called **rumination**.

Not-So-Fun Fact!
If an annoyed camel spits at you, it is not spit but brought-up stomach stuff! Yuck.

Answers: (1) Foramen (2) Temporal bone

5

The Skull

 Color these skulls with the same colors you used for the dog skull. Use the numbers to identify and choose the right color for each bone.

Mammals that bang heads with each other often have horns or antlers projecting from the skull, so the bones that attach to them are larger than on other mammals.

Rhinoceroses have horns on the nose, which is why the **nasal bone** (2) is large and strong.

Buffalo horns are made of the same substance as fingernails and grow from the skin on **projections** (5a) of their large frontal bone.

Deer antlers are bones too. They just aren't covered by skin.

Did you know...
that rhino horn is not horn, but has a structure like compacted hair?

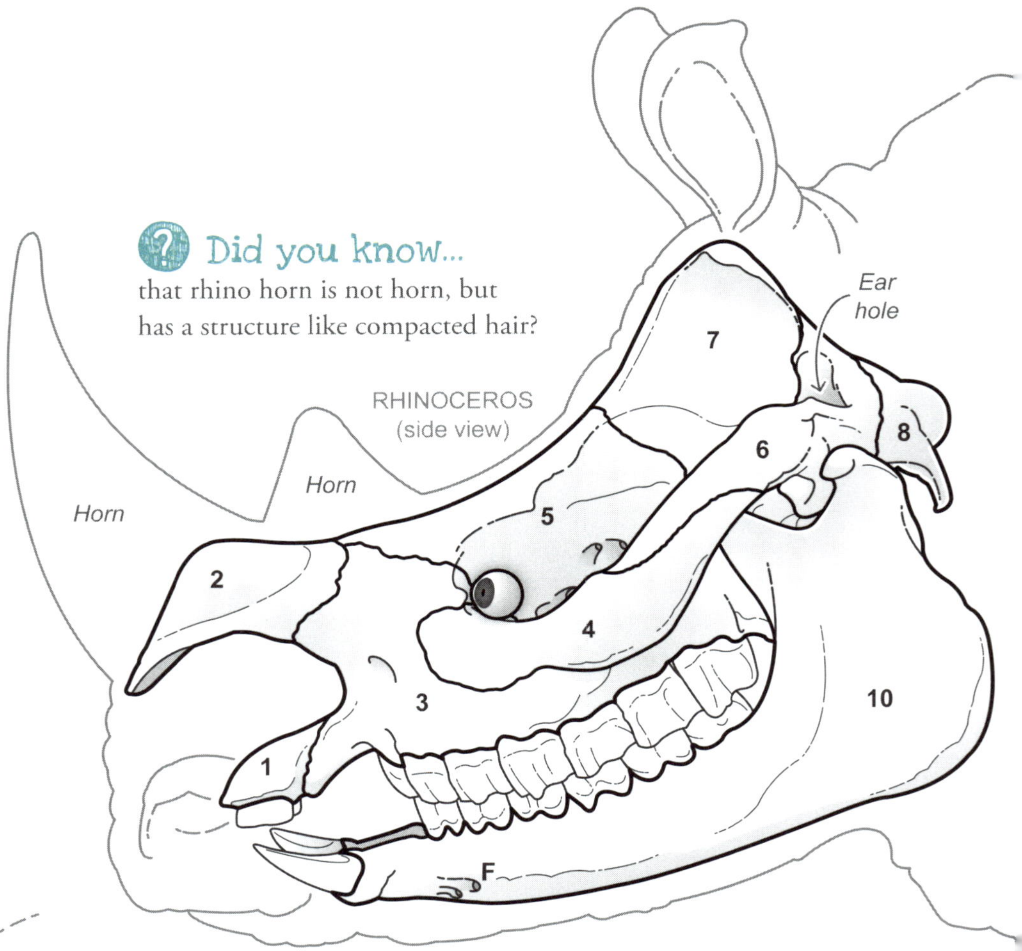

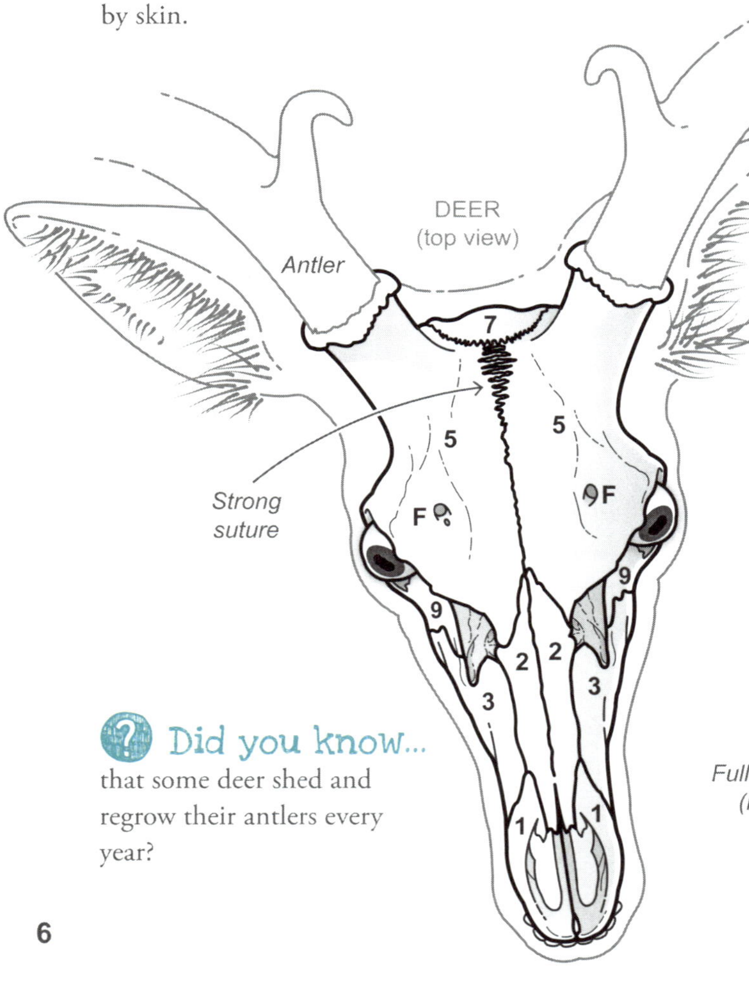

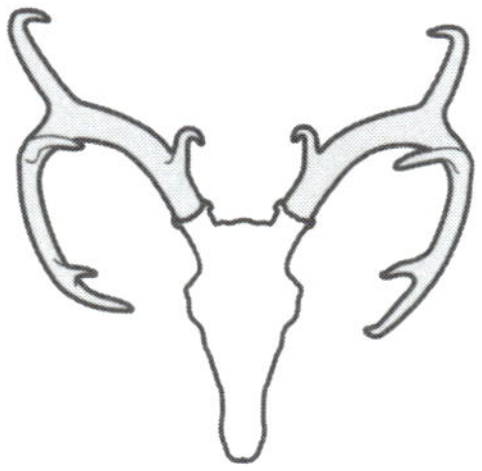

RHINOCEROS (side view)

DEER (top view)

Strong suture

Did you know...
that some deer shed and regrow their antlers every year?

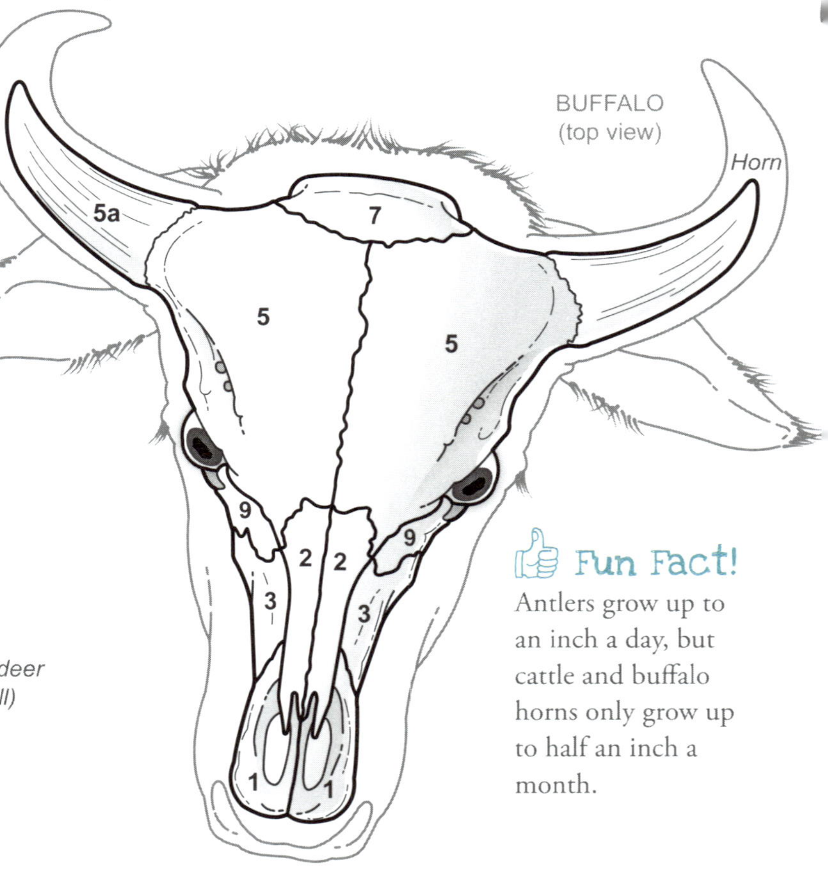

Full size of antlers on this deer (in proportion to the skull)

BUFFALO (top view)

Fun Fact!
Antlers grow up to an inch a day, but cattle and buffalo horns only grow up to half an inch a month.

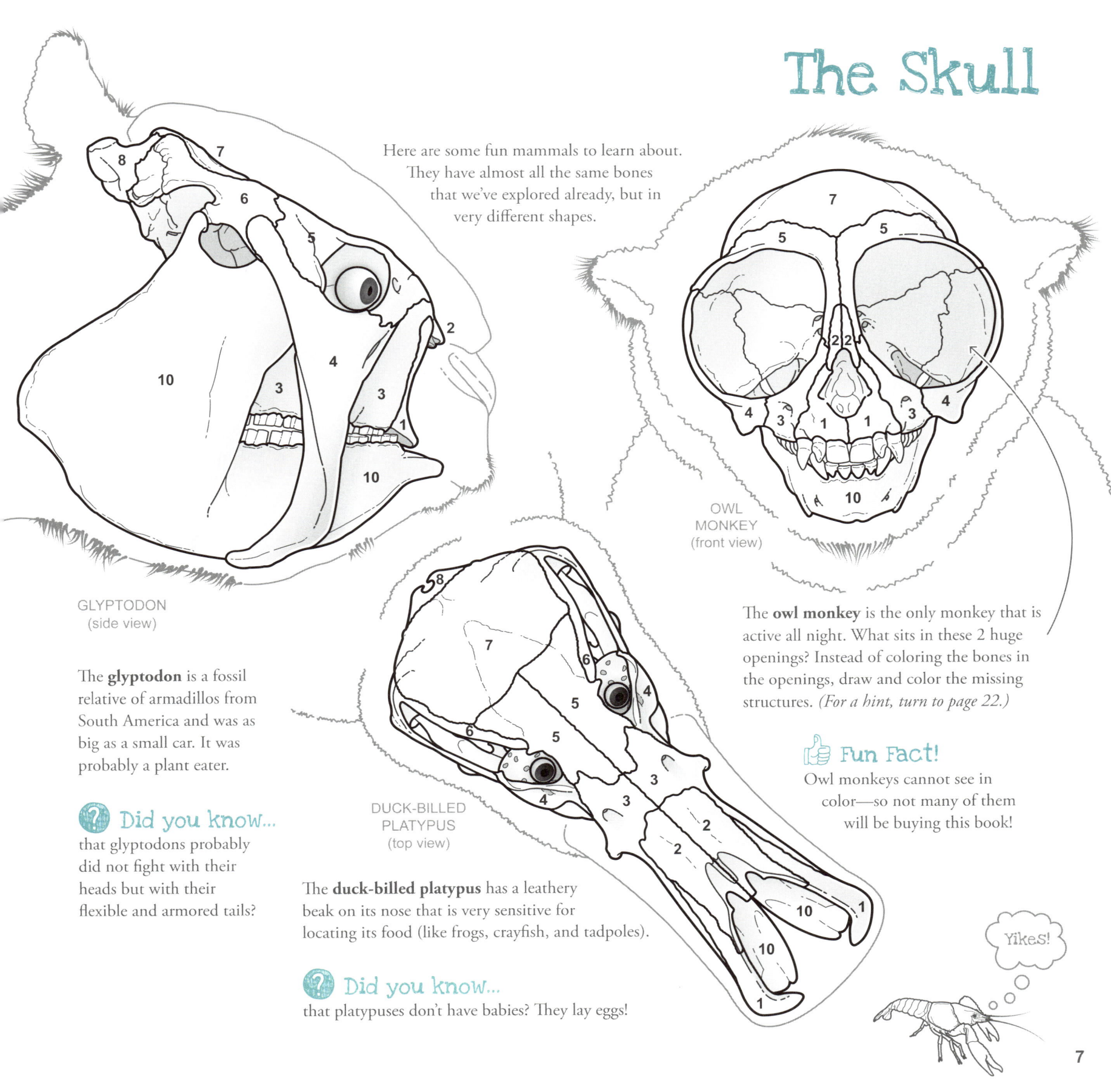

The Skull

Here are some fun mammals to learn about. They have almost all the same bones that we've explored already, but in very different shapes.

GLYPTODON
(side view)

The **glyptodon** is a fossil relative of armadillos from South America and was as big as a small car. It was probably a plant eater.

❓ Did you know...
that glyptodons probably did not fight with their heads but with their flexible and armored tails?

DUCK-BILLED PLATYPUS
(top view)

The **duck-billed platypus** has a leathery beak on its nose that is very sensitive for locating its food (like frogs, crayfish, and tadpoles).

❓ Did you know...
that platypuses don't have babies? They lay eggs!

OWL MONKEY
(front view)

The **owl monkey** is the only monkey that is active all night. What sits in these 2 huge openings? Instead of coloring the bones in the openings, draw and color the missing structures. *(For a hint, turn to page 22.)*

👍 Fun Fact!
Owl monkeys cannot see in color—so not many of them will be buying this book!

Yikes!

7

Arms on Land

In the first chapter, we showed you how skulls, the bones that make them up, and the teeth are all different sizes and shapes depending on what an animal eats.

When we look at their arms (front legs or forelimbs), it's clear that the shape and structure of the arm changes based on how an animal moves—either on land, in the air, or in water. The animals in this chapter have many of the same bones in their arms, but arms for running, flying, and swimming all look different.

 Notice that each arm bone is identified by a specific number. Shade bones with the same number in the same color on every drawing in this chapter. Then test your veterinarian to see if they know the bones as well as you do!

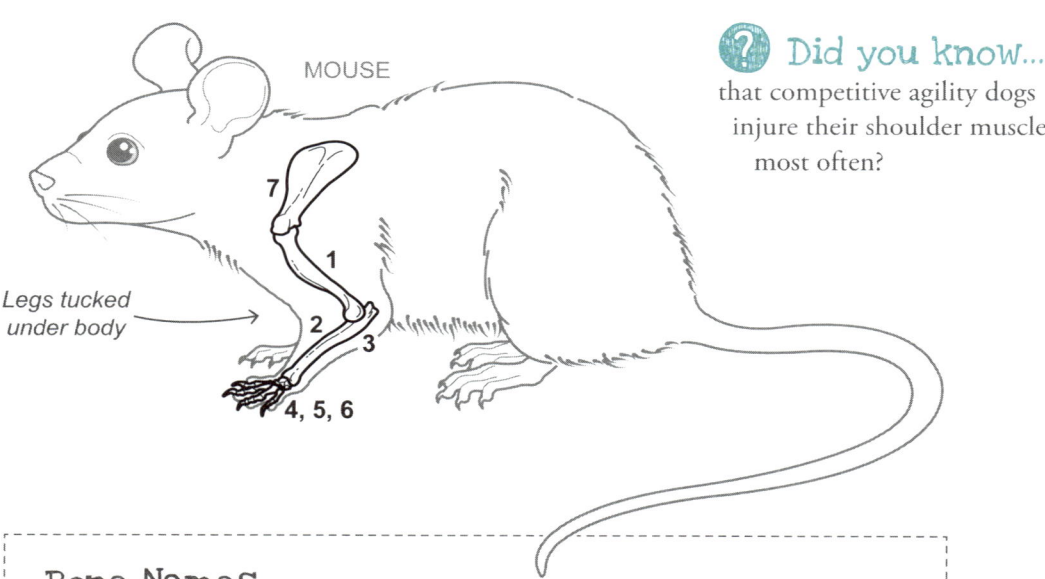

MOUSE

Legs tucked under body

? Did you know...
that competitive agility dogs injure their shoulder muscles most often?

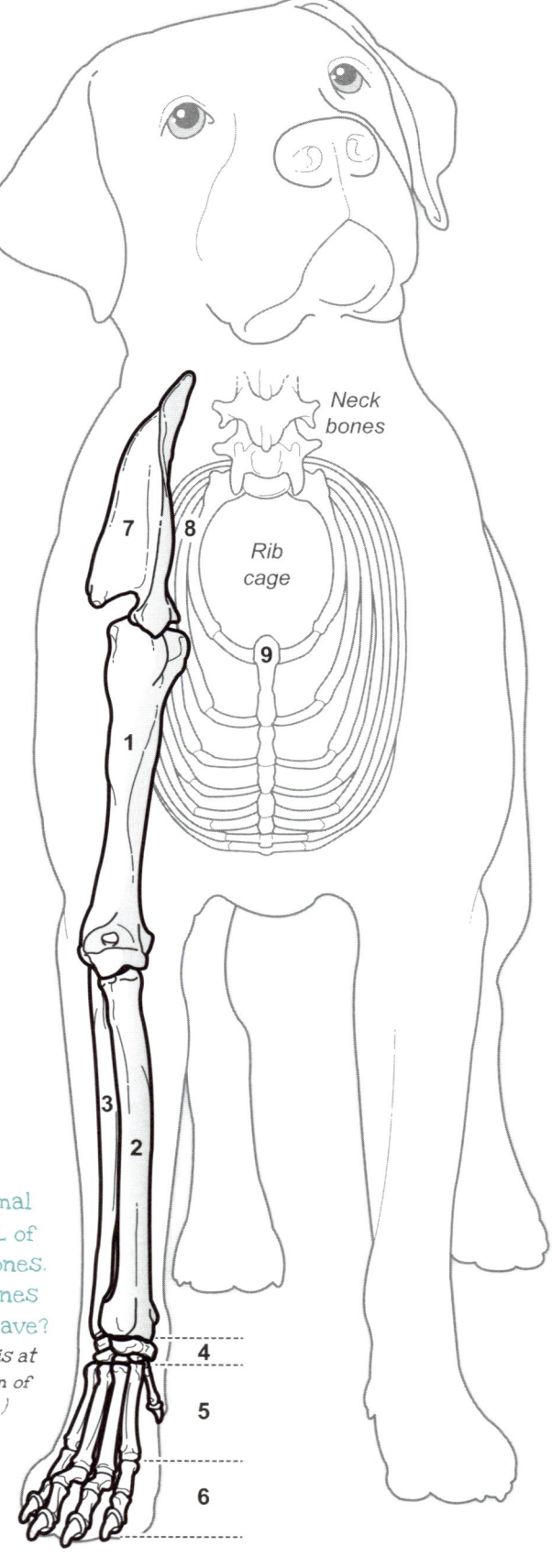

Neck bones

Rib cage

No animal has ALL of these bones. Which ones do you have?
(Answer is at the bottom of the page.)

Bone Names

These are the arm bones:
1. Humerus (upper arm)
2. Radius ⎤
3. Ulna ⎦ (forearm)
4. Carpal bones (wrist)
5. Metacarpal bones (palm)
6. Phalanges (fingers)

These bones support the arm on the body:
7. Scapula (shoulder blade)
8. Ribs
9. Sternum (breast bone or bones)
10. Clavicle (collarbone)
11. Furcula (wishbone)
12. Coracoid (that's its name!)

Answer: You have bones numbered 1-10.

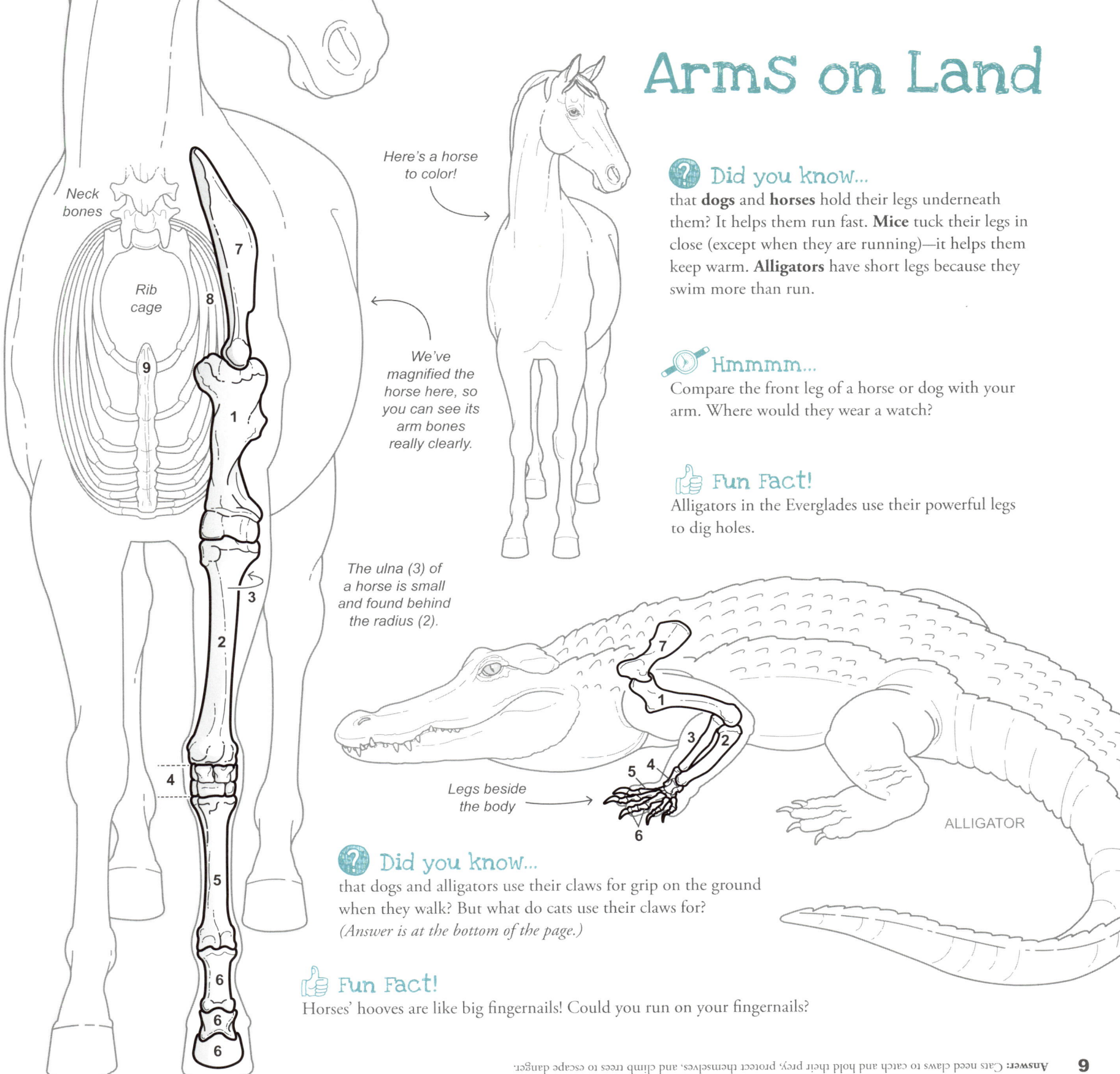

Arms in the Air

✏️ Use the "Bone Names" list on the previous page to identify and color the wing bones and supporting bones. Birds and bats have some bones that are not found in horses and dogs. Which ones are they?

Wouldn't you love to be able to fly? Your arms would have to look very different for you to do that—they'd have to be wings. **Bird and bat wings have mostly the same arm bones as us, but in each animal, those bones are turned into wings in different ways.** Birds fuse most of the hand bones—**carpals** (4), **metacarpals** (5), and **phalanges** (6)—together to form a strong base for feathers, which are specialized scales. Bats elongate the **metacarpals** (5) and **phalanges** (6) and stretch the skin between them to make a sail-like wing.

❓ **Did you know…**
that the **furcula** (11) is a spring that helps the bird move its wings up and down? The strong **coracoid** (12) anchors the wing on the **ribs** (8) and huge **sternum** (9).

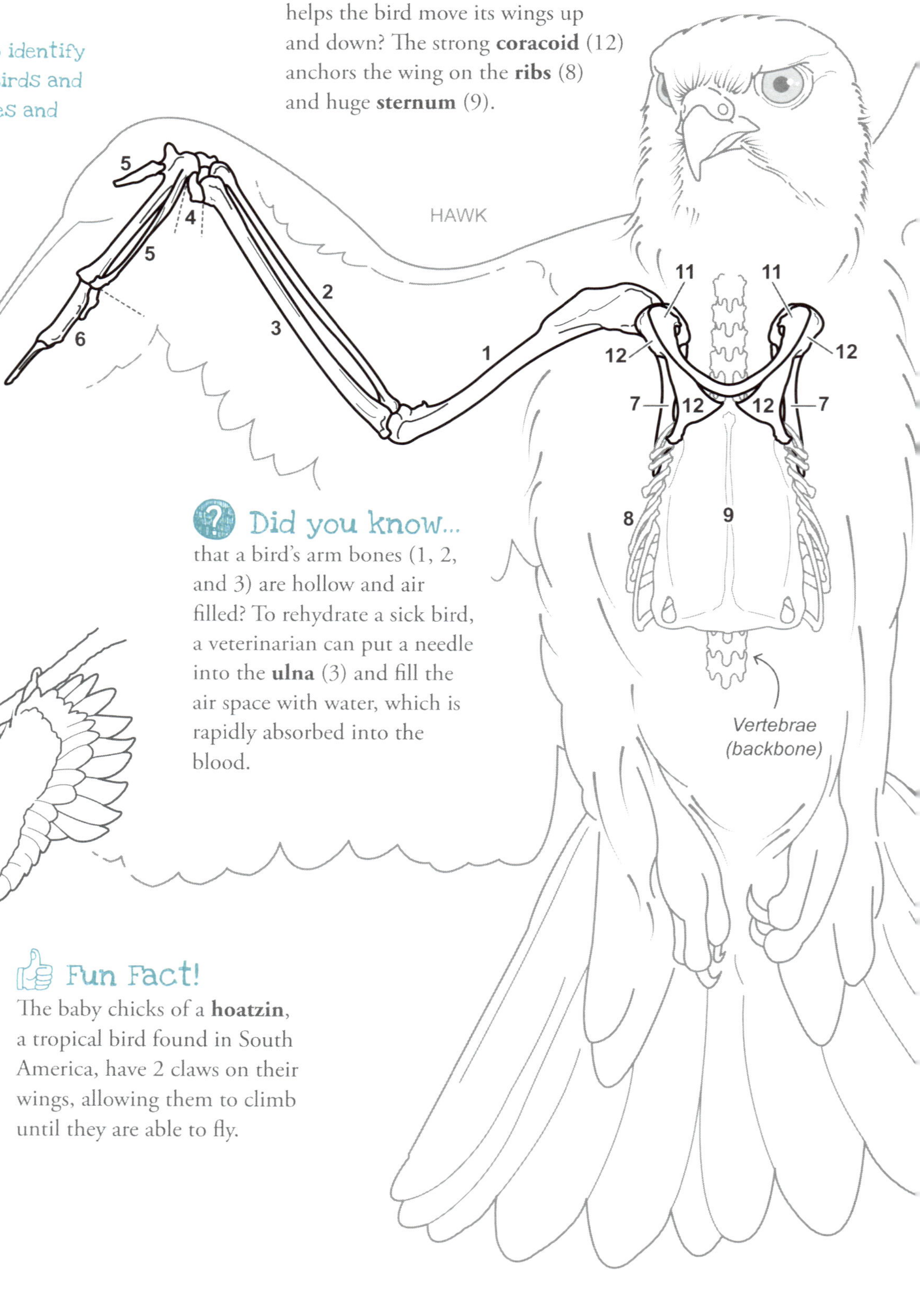

❓ **Did you know…**
that a bird's arm bones (1, 2, and 3) are hollow and air filled? To rehydrate a sick bird, a veterinarian can put a needle into the **ulna** (3) and fill the air space with water, which is rapidly absorbed into the blood.

👍 **Fun Fact!**
The baby chicks of a **hoatzin**, a tropical bird found in South America, have 2 claws on their wings, allowing them to climb until they are able to fly.

HAWK

Vertebrae (backbone)

HOATZIN

Claws on wing

10

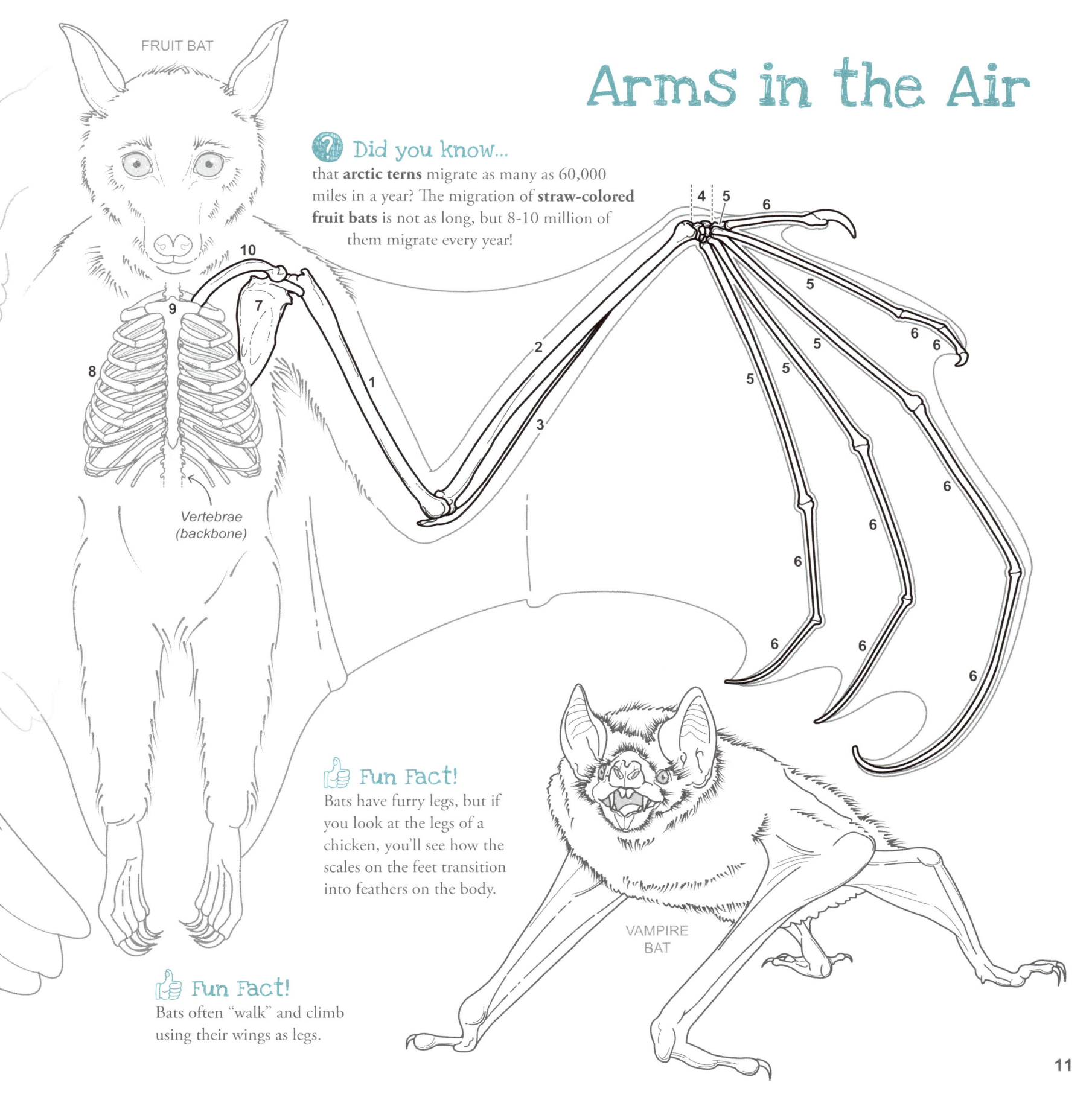

Arms in the Water

Animals that live on land and in the air use their arms (and legs) when they move around. **But for animals that live in the water, the tail is the main structure that provides the thrust for moving by swimming. Arms and legs become fins or flippers that are mostly used for steering.** Dolphin flippers have bones and muscles, but fish fins are made of stiff rays with a membrane (a bit like a bat wing) and only a few bones.

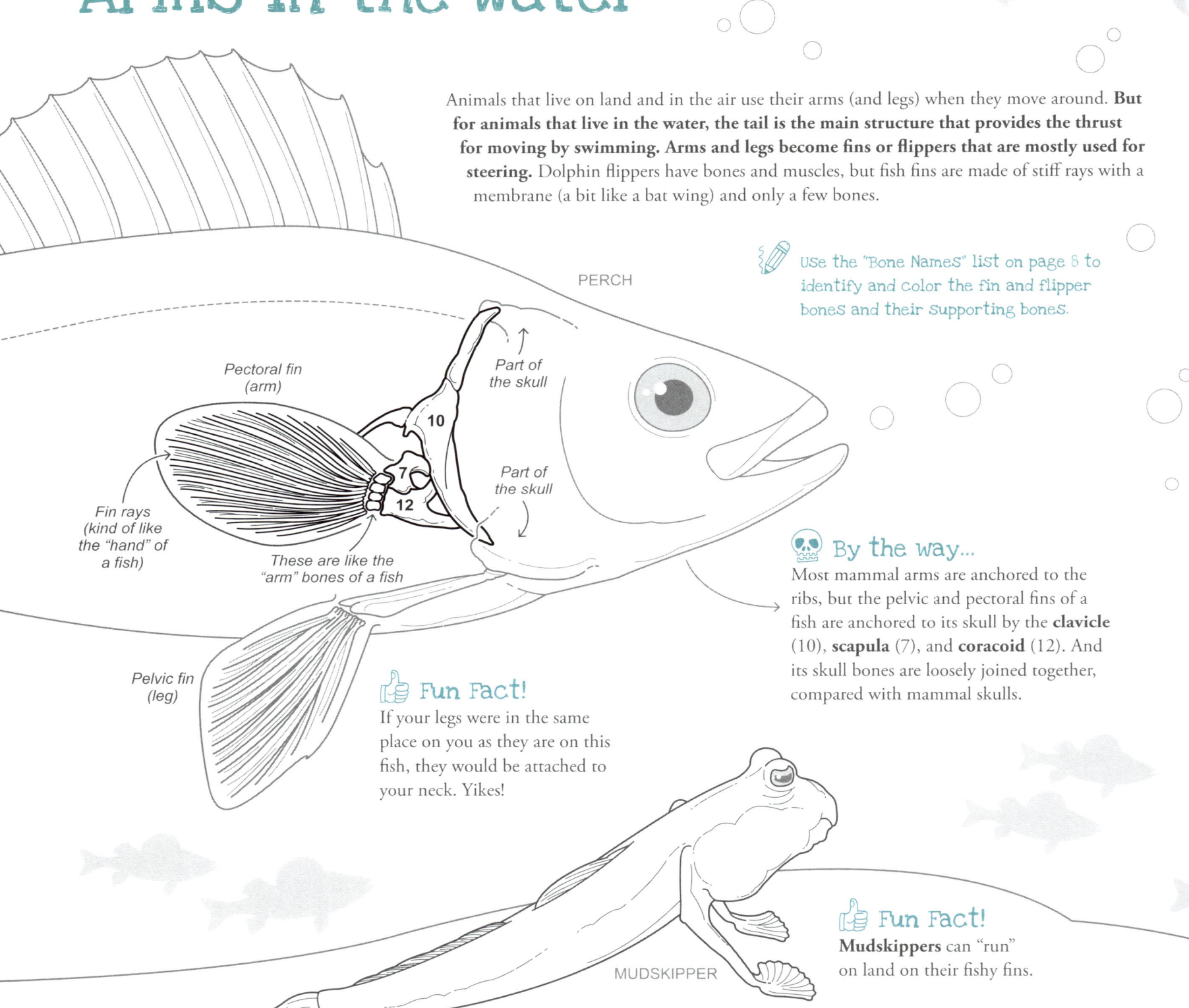

PERCH

Use the "Bone Names" list on page 8 to identify and color the fin and flipper bones and their supporting bones.

Pectoral fin (arm)

Part of the skull

Part of the skull

Fin rays (kind of like the "hand" of a fish)

These are like the "arm" bones of a fish

Pelvic fin (leg)

By the way...
Most mammal arms are anchored to the ribs, but the pelvic and pectoral fins of a fish are anchored to its skull by the **clavicle** (10), **scapula** (7), and **coracoid** (12). And its skull bones are loosely joined together, compared with mammal skulls.

Fun Fact!
If your legs were in the same place on you as they are on this fish, they would be attached to your neck. Yikes!

MUDSKIPPER

Fun Fact!
Mudskippers can "run" on land on their fishy fins.

Arms in the Water

🛈 Did you know...
that **dolphins** tend to lose the leg and hip (pelvic) bones altogether, while many fish have their legs in front of their arms, to give 2 sets of steering fins near the head?

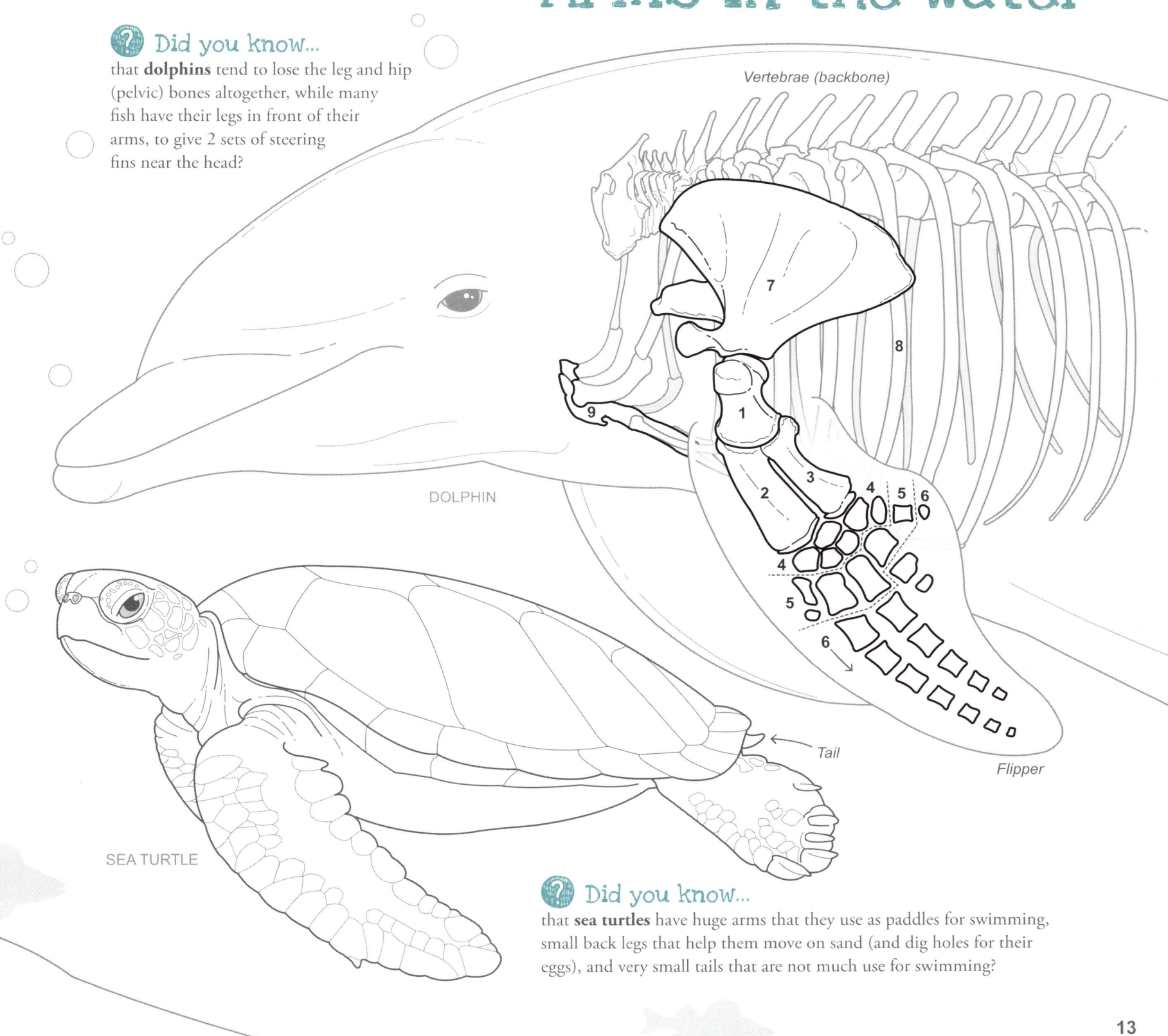

DOLPHIN

SEA TURTLE

Vertebrae (backbone)

Tail

Flipper

🛈 Did you know...
that **sea turtles** have huge arms that they use as paddles for swimming, small back legs that help them move on sand (and dig holes for their eggs), and very small tails that are not much use for swimming?

Eating & Digesting on Land

All animals eat and digest food so they can get the nutrients and energy they need to grow, keep healthy, and be active. This requires lots of different organs, which together are called the **digestive system**. It's a complicated system because many processes are involved between the time food enters the mouth and when feces (poop) exits the body at the anus.

We've stretched out the dog's digestive system here to show you how each organ connects with another, but in reality, most of the organs are all bunched up neatly together in the **abdomen** (belly). The blue-green arrows indicate which direction the food travels in.

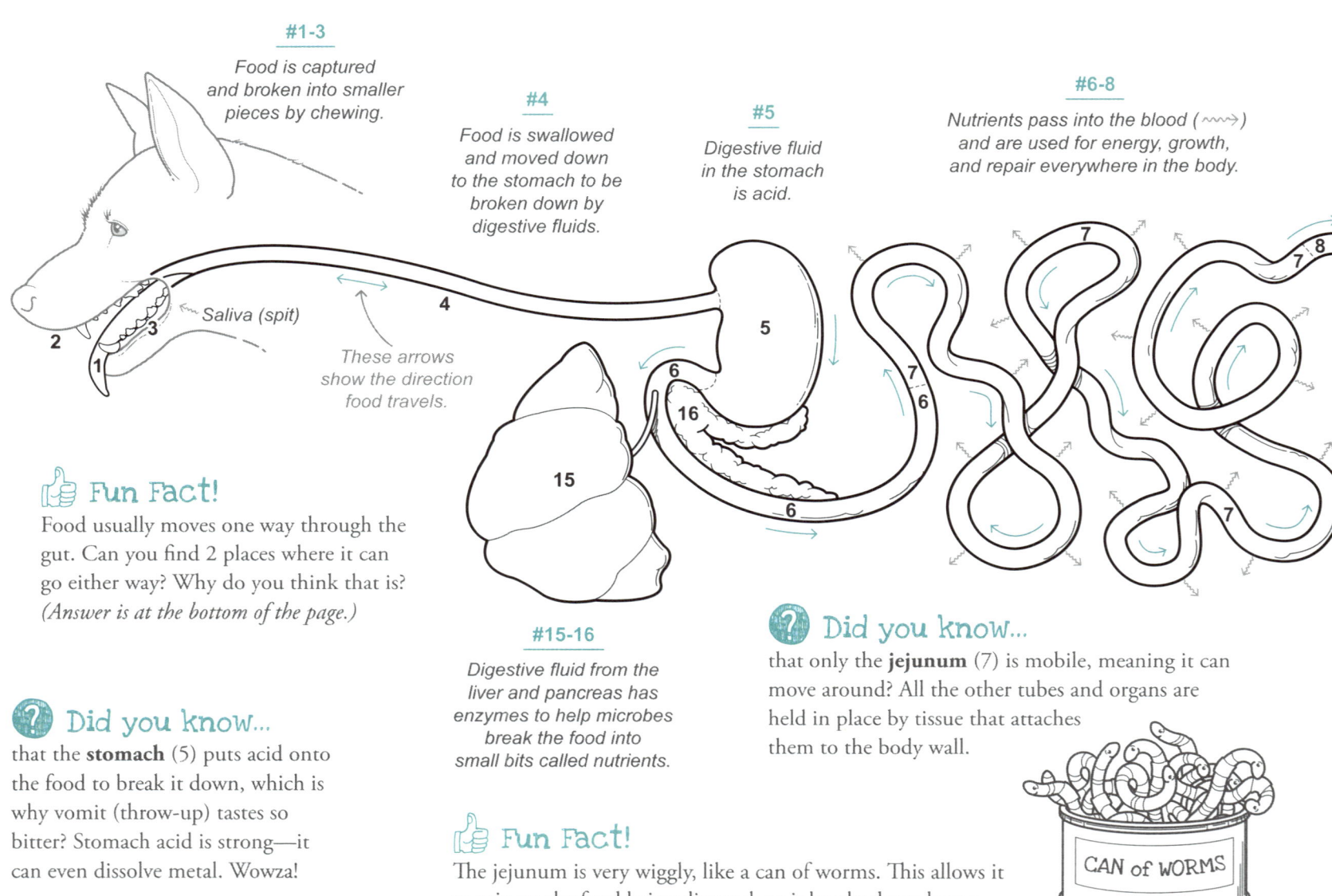

#1-3
Food is captured and broken into smaller pieces by chewing.

#4
Food is swallowed and moved down to the stomach to be broken down by digestive fluids.

#5
Digestive fluid in the stomach is acid.

#6-8
Nutrients pass into the blood (⤳) and are used for energy, growth, and repair everywhere in the body.

These arrows show the direction food travels.

👍 **Fun Fact!**
Food usually moves one way through the gut. Can you find 2 places where it can go either way? Why do you think that is? *(Answer is at the bottom of the page.)*

❓ **Did you know...**
that the **stomach** (5) puts acid onto the food to break it down, which is why vomit (throw-up) tastes so bitter? Stomach acid is strong—it can even dissolve metal. Wowza!

#15-16
Digestive fluid from the liver and pancreas has enzymes to help microbes break the food into small bits called nutrients.

👍 **Fun Fact!**
The jejunum is very wiggly, like a can of worms. This allows it to mix up the food being digested, so it breaks down better.

❓ **Did you know...**
that only the **jejunum** (7) is mobile, meaning it can move around? All the other tubes and organs are held in place by tissue that attaches them to the body wall.

Answer: If you swallow something nasty, vomiting brings food back up from the stomach. Digested food goes in and out of the cecum through its 1 opening.

The Meat Eaters

Each part of the digestive system has a different role to play. It has chambers where food stays a while (5, 9), long tubes that take time for food to get through (6-8, 10), and organs that add substances called "enzymes" that help break down the food (15, 16).

✏️ Each organ in the digestive system is identified by a specific number. Shade parts with the same number in the same color on every drawing in this chapter. If you do this, you'll be amazed at how much the digestive tracts of various animals differ!

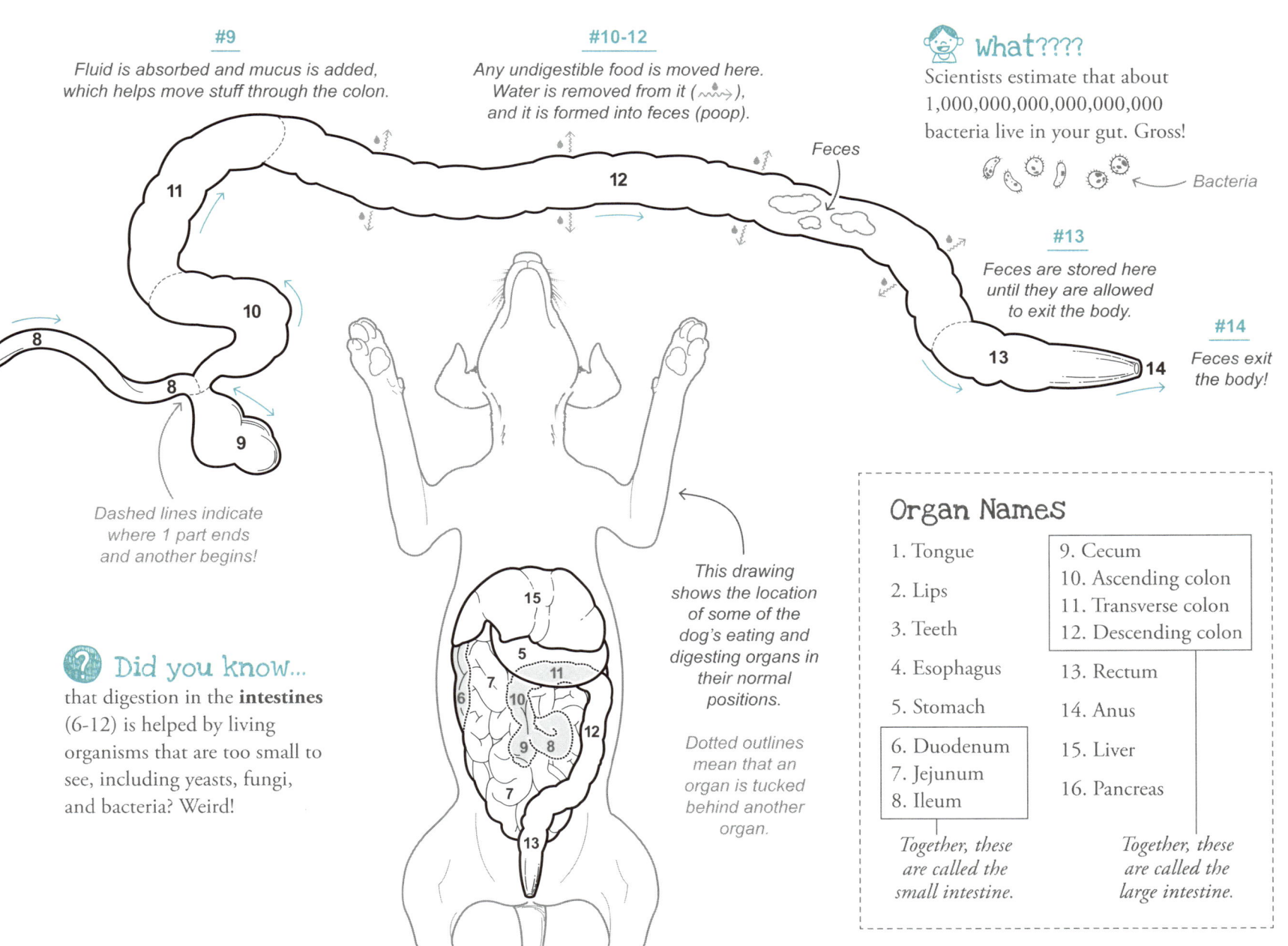

#9
Fluid is absorbed and mucus is added, which helps move stuff through the colon.

#10-12
Any undigestible food is moved here. Water is removed from it ( ), and it is formed into feces (poop).

what????
Scientists estimate that about 1,000,000,000,000,000,000 bacteria live in your gut. Gross!

#13
Feces are stored here until they are allowed to exit the body.

#14
Feces exit the body!

Dashed lines indicate where 1 part ends and another begins!

This drawing shows the location of some of the dog's eating and digesting organs in their normal positions.

Dotted outlines mean that an organ is tucked behind another organ.

❓ Did you know...
that digestion in the **intestines** (6-12) is helped by living organisms that are too small to see, including yeasts, fungi, and bacteria? Weird!

Organ Names

1. Tongue
2. Lips
3. Teeth
4. Esophagus
5. Stomach
6. Duodenum
7. Jejunum
8. Ileum

9. Cecum
10. Ascending colon
11. Transverse colon
12. Descending colon
13. Rectum
14. Anus
15. Liver
16. Pancreas

Together, these (6, 7, 8) are called the small intestine.

Together, these (9, 10, 11, 12) are called the large intestine.

15

Eating & Digesting on Land

Plants are more difficult to digest than meat, so grass eaters like horses and cows rely on a process called **fermentation** to help break down food. Fermentation is the digestion of food using fluid and microorganisms. Meat eaters, like cats, use fermentation in their intestines too, but not nearly as much as grass eaters—who have much, much bigger fermentation organs in their digestive systems. Like the dog, we've stretched out the digestive system of the cow and horse so you can see all the organs clearly.

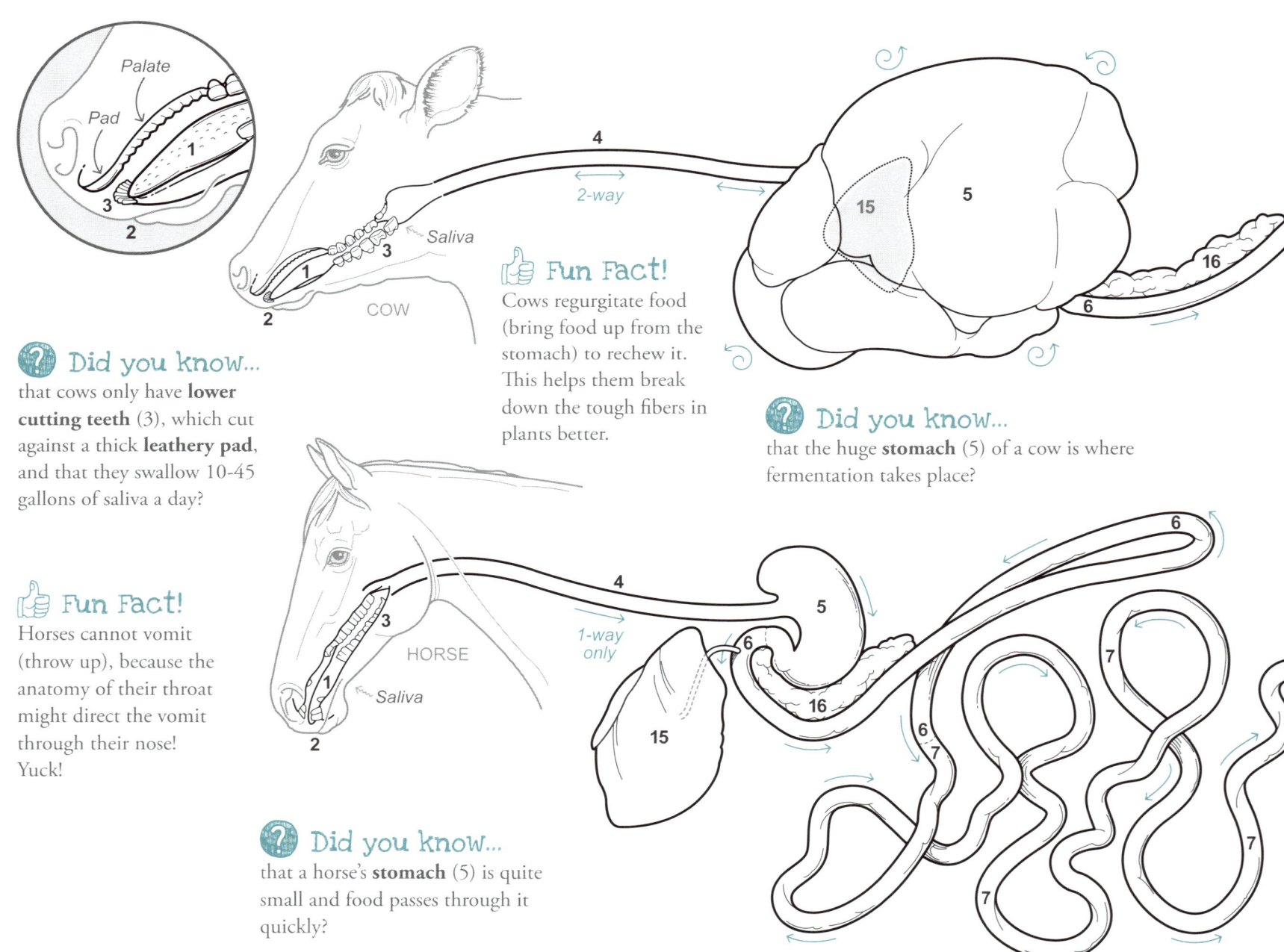

👍 **Fun Fact!**
Cows regurgitate food (bring food up from the stomach) to rechew it. This helps them break down the tough fibers in plants better.

❓ **Did you know...**
that cows only have **lower cutting teeth** (3), which cut against a thick **leathery pad**, and that they swallow 10-45 gallons of saliva a day?

❓ **Did you know...**
that the huge **stomach** (5) of a cow is where fermentation takes place?

👍 **Fun Fact!**
Horses cannot vomit (throw up), because the anatomy of their throat might direct the vomit through their nose! Yuck!

❓ **Did you know...**
that a horse's **stomach** (5) is quite small and food passes through it quickly?

The Grass Eaters

 Continue to shade organs with the same number in the same color. Many people (including adults!) think horses and cows have pretty similar digestive systems, but as you color, you'll notice how massive a cow stomach is when compared to a horse!

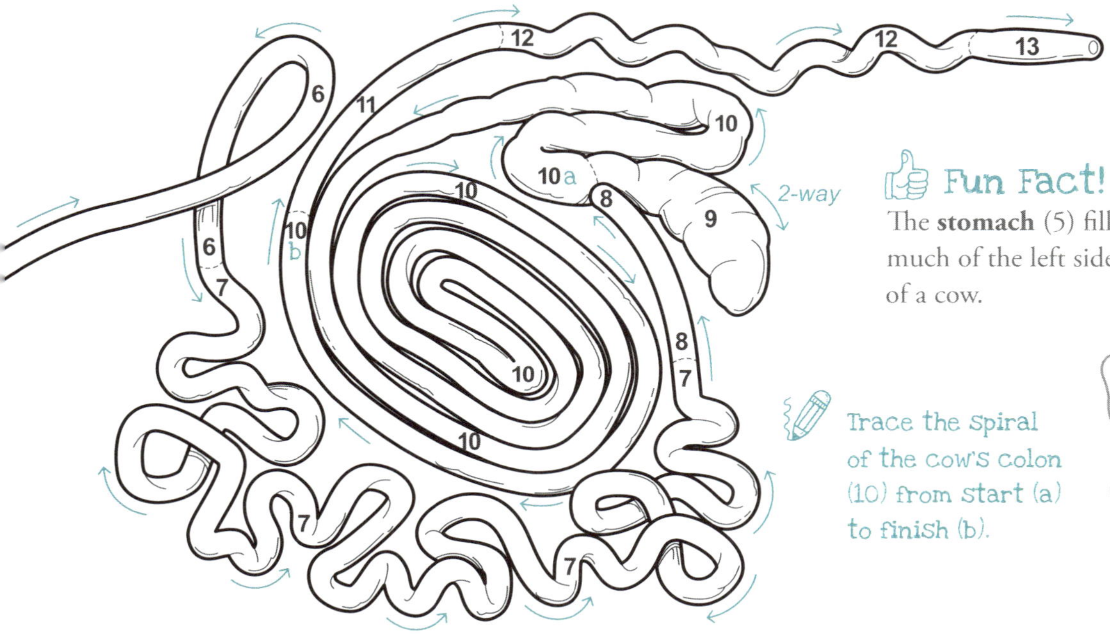

Left side view of the cow showing the size and location of the stomach (5)

👍 Fun Fact!
The **stomach** (5) fills much of the left side of a cow.

✏️ Trace the spiral of the cow's colon (10) from start (a) to finish (b).

❓ Did you know...
that the **colon** (10) is enlarged in the horse and acts like a huge fermentation chamber?

👍 Fun Fact!
The **colon** (10) fills much of the lower half of the belly of a horse.

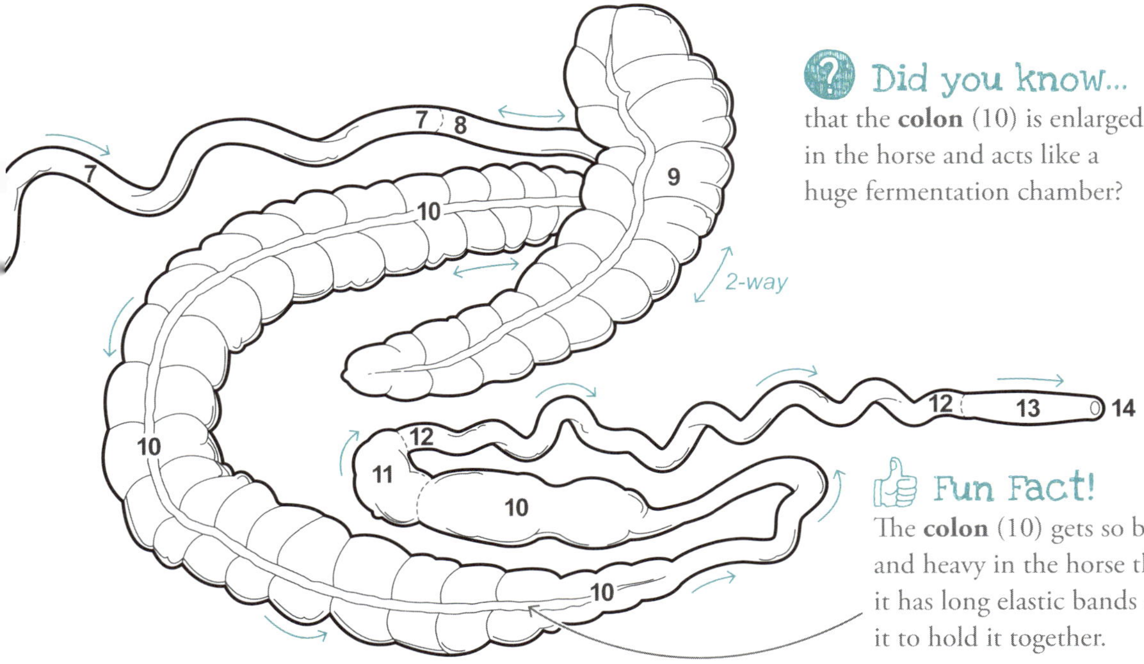

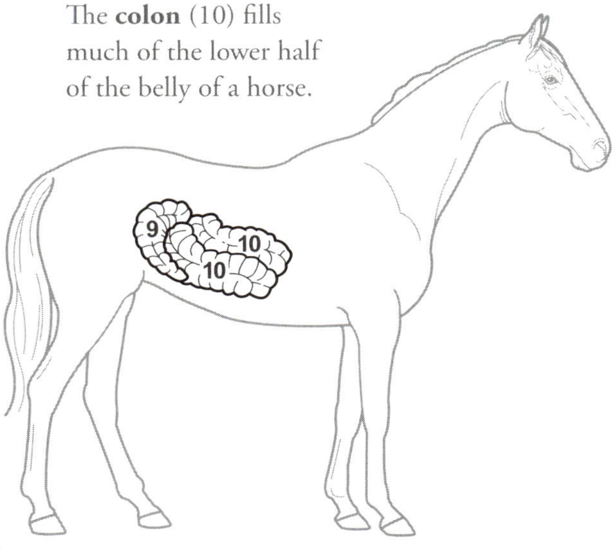

👍 Fun Fact!
The **colon** (10) gets so big and heavy in the horse that it has long elastic bands on it to hold it together.

Right side view of the horse showing the size and location of the cecum (9) and colon (10)

17

Eating & Digesting in Air

Birds and fish have digestive systems that are similar to those of land animals. They have a mouth at the front, an exit at the other end, and various tubes, organs, and chambers in between. BUT their parts are not exactly the same as dogs, cows, horses, and other mammals.

Here are the digestive systems of a bird and a shark.

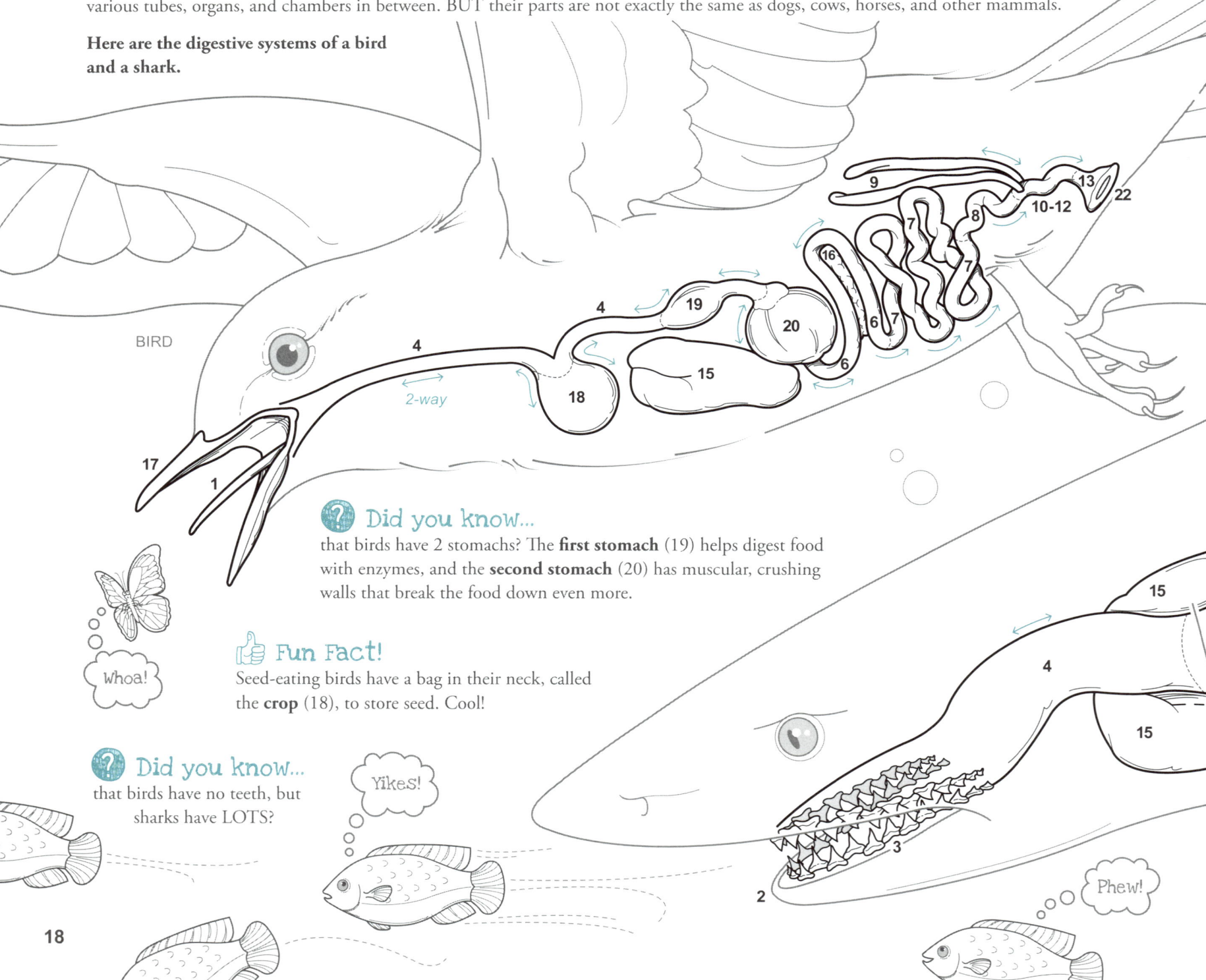

❓ Did you know...
that birds have 2 stomachs? The **first stomach** (19) helps digest food with enzymes, and the **second stomach** (20) has muscular, crushing walls that break the food down even more.

👍 Fun Fact!
Seed-eating birds have a bag in their neck, called the **crop** (18), to store seed. Cool!

❓ Did you know...
that birds have no teeth, but sharks have LOTS?

Eating & Digesting in Water

Continue to shade organs with the same number in the same color. You'll notice that birds have several organs that differ from the other animals in this chapter—such as the beak. Can you name other animals that have beaks? *(Answer is at the bottom of the page.)*

Organ Names

17. Beak
18. Crop
19. Proventriculus (first stomach)
20. Gizzard (second stomach)
21. Rectal gland
22. Cloaca (anus)

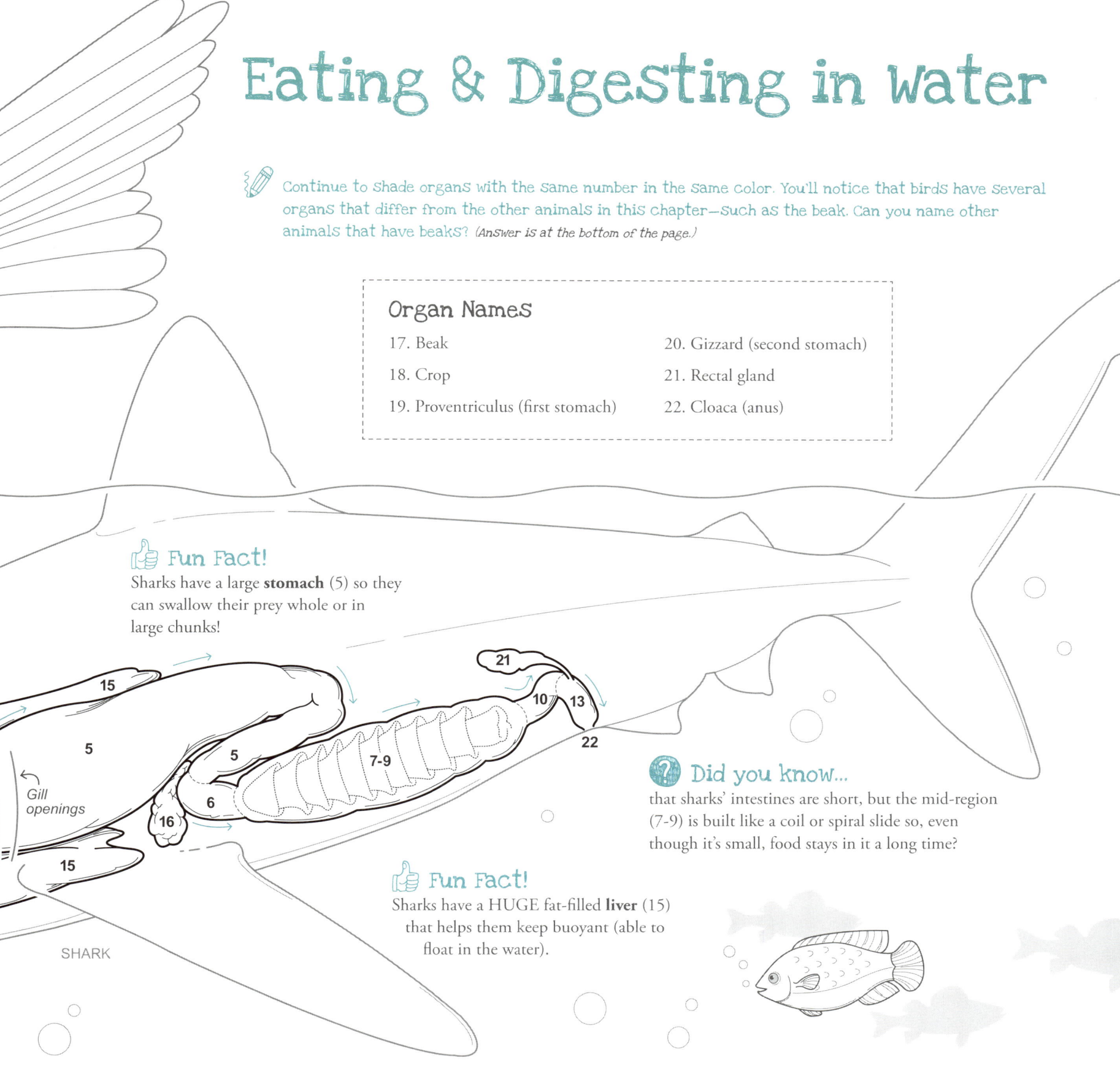

👍 Fun Fact!
Sharks have a large **stomach** (5) so they can swallow their prey whole or in large chunks!

❓ Did you know...
that sharks' intestines are short, but the mid-region (7-9) is built like a coil or spiral slide so, even though it's small, food stays in it a long time?

👍 Fun Fact!
Sharks have a HUGE fat-filled **liver** (15) that helps them keep buoyant (able to float in the water).

Answer: Turtle, squid, octopus, cuttlefish, duck-billed platypus, and echidna (they look like a cross between an anteater and a porcupine).

The Eye

Animals have specialized structures—eyes and ears—for detecting sights and sounds in their environments. For many animals, these organs can be particularly important for finding food or avoiding becoming someone else's dinner.

We'll explore eyes first and then ears. The eyes of most mammals are very similar in structure.

✏️ Shade the eye parts with the same number in the same color on every eye drawing in this chapter.

👍 **Fun Fact!**
The eye is like a video camera, continually taking moving pictures.

Inside structure of the eye

👍 **Fun Fact!**
The picture that the eye sees is upside down. The brain turns it right way up!

Parts of the Retina

Layers of cells in the retina

13. **Rods:** Cells very sensitive to white light

14. **Cones:** Cells sensitive to color—each cone "sees" 1 color

15. **Other** cells and branches of the optic nerve

❓ **Did you know...**
that the eye is moved by 7 muscles? Four **straight muscles** (S) move it side to side and up and down, 2 **oblique muscles** (O) tilt it inward or outward, and 1 small **retractor muscle** (R) pulls it back into the socket a little way (humans don't have this muscle). The retractor muscle allows the nictitating membrane to wipe the eye without getting hung up on the eyelids!

This muscle lifts the eyelid!

Side view of eye and eye muscles

Parts of the Eye & What They Do

1. **Cornea:** "Clear window" that lets light into the eye
2. **Two fluid-filled chambers:** These help the eye keep its shape
3. **Pupil:** Hole that lets light into the inner chamber
4. **Iris:** Colored membrane that changes the size of the pupil
5. **Lens:** Focuses images on the back of the eye
6. **Ciliary muscle:** Changes the lens shape for near and far focus
7. **Fibers:** "Strings" that connect the lens to the ciliary muscle
8. **Sclera:** Tough white coating that protects the eye
9. **Choroid:** Black layer that stops light bouncing around in the eye
10. **Tapetum:** Colored mirror that improves low-light vision ← *We don't have this!*
11. **Retina:** Converts light to nerve impulses
12. **Optic nerve:** Takes nerve impulses to the brain

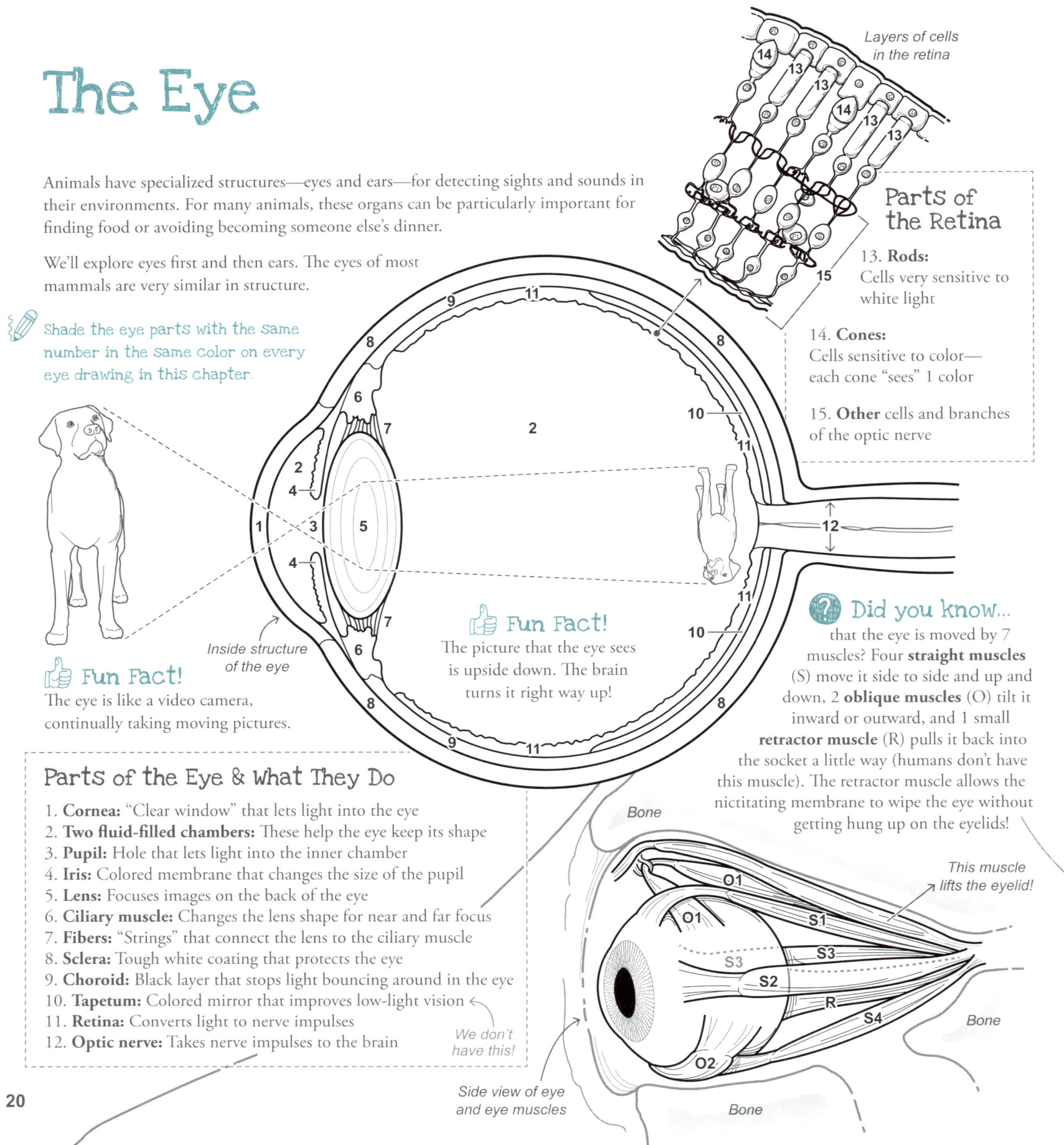

Eyes of Predators

What we notice most about eyes are the color of the **iris** (4) and the shape of the **pupil** (3). Let's explore the eyes of predators that eat meat...

 Did you know...
that animals that ambush (jump out) at their prey (like **cats** do) often have vertical slit pupils, because this helps them judge how far away their dinner is?

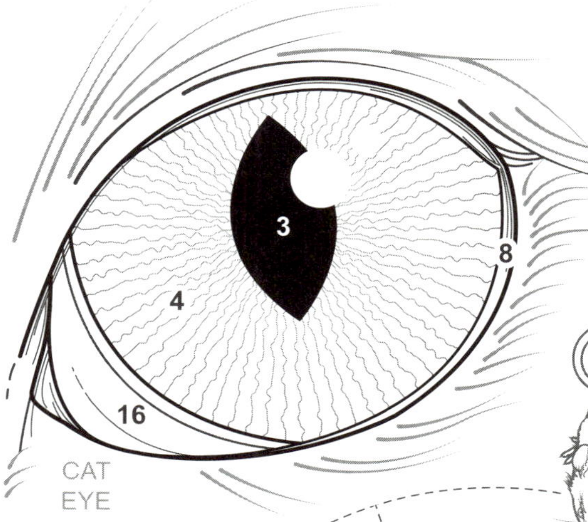

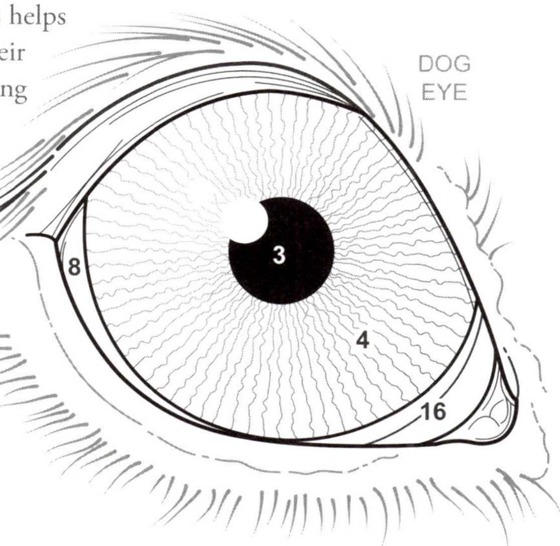

👍 Fun Fact!
Tigers can see 8 times better than us at night because they have many more **rods** (13) in the retina.

 Did you know...
that meat eaters often have their eyes facing forward (as we do)? This is called **binocular vision**.
Seeing prey with both eyes also helps to judge distances— "How far do I need to jump to catch my dinner?"

 Did you know...
that many animals have a membrane that can wipe across the eye to clean it, like windshield wipers on a car? It has the fun name of **nictitating membrane** (16). We don't have one.

 Did you know...
that animals that hunt (chase after) their prey, like **wild dogs** and **wolves**, often have round pupils, because this helps them see their prey over long distances?

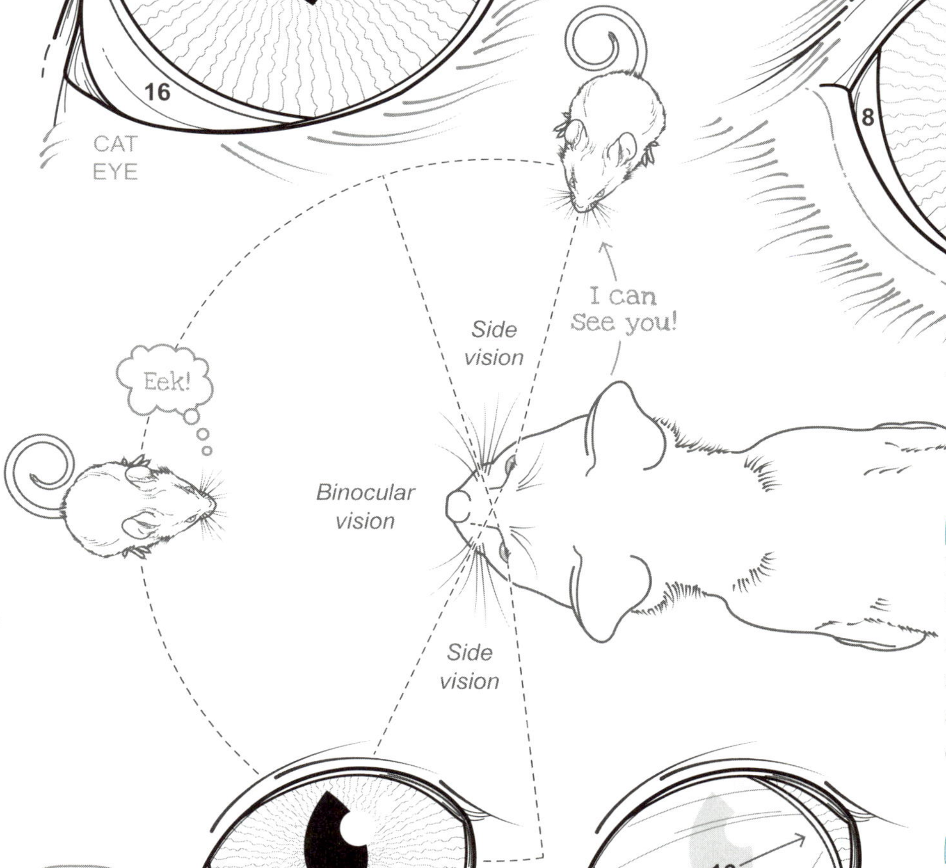

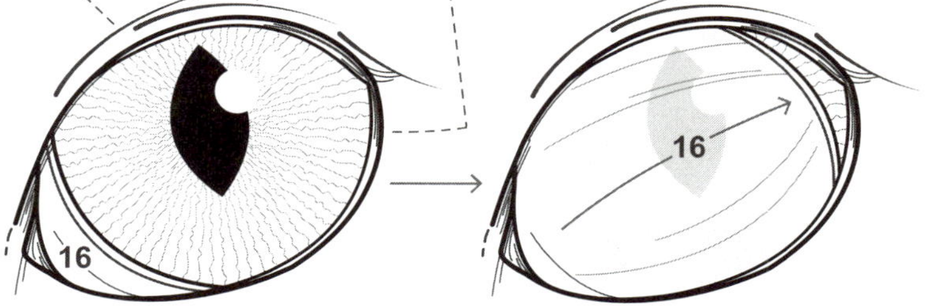

👍 Fun Fact!
Even if the eyes face forward, being able to move them with the eye muscles means that the animal can still see "out of the corner of its eye."

👍 Fun Fact!
The nictitating membrane of a **polar bear** filters out harmful ultraviolet (UV) radiation.

CAT EYE

DOG EYE

I can see you!

Eek!

Side vision

Binocular vision

Side vision

Nictitating membrane retracted (pulled back)

Nictitating membrane extended over the eye

Eyes of Prey Animals

On this page, you'll see how the eyes of prey animals (animals that are food for other animals) differ from the eyes of predators. The most obvious differences are pupil shape and where the eyes are positioned on the head. You'll meet an old friend from the skull chapter—the owl monkey—and a strange new one—the chameleon.

Did you know...
that the pupils of prey animals are often horizontal slits? This allows the animal to scan more of the horizon for danger. A slit pupil can expand to let more light in than a round one.

Fun Fact!
Horses have little tags dangling from their iris. These may act as mini sunshades and also help to detect predators as they appear to flicker in and out of view between the tags.

Did you know...
that prey animals often have eyes on the sides of the head? This is called **monocular vision**, and it allows them to see almost all the way around their bodies. Imagine that! No one would ever be able to sneak up on you!

Fun Fact!
Owl monkeys have huge eyes with round pupils to let more light in at night. They have few cones in their retinas, so they do not see in color.

Fun Fact!
Chameleon eyes don't move together. Each eye rotates independently of the other eye. I wonder what their world looks like?

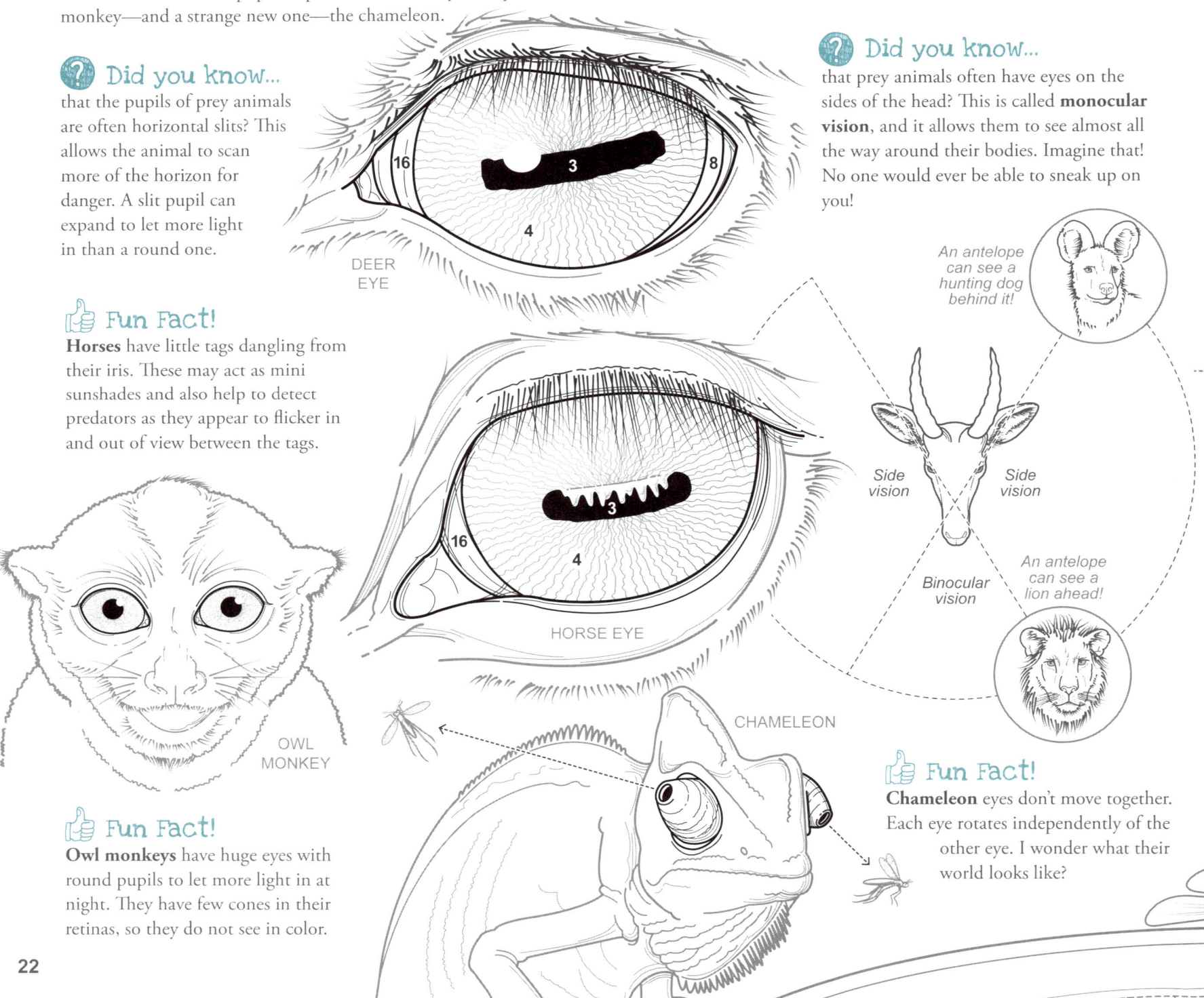

Eyes in Air & Water

The eyes of animals that fly and swim have similar anatomy to those of the animals you've seen so far, but there are some interesting differences. Most birds have long rather than round eyes, and many have monocular vision. But the hunters—owls, hawks, and eagles—all have eyes forward, which gives them excellent binocular vision. Most fish have a similar eye structure to mammals, but some lack eye muscles, and all bony fish have a fixed pupil size.

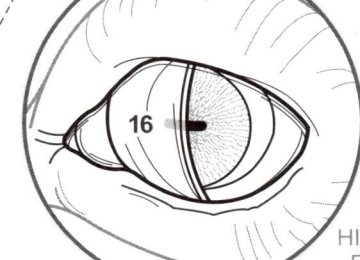

OWL

Special Parts of Bird Eyes

17. Sclerotic ring: Ring of bone in the wall of the eyeball that supports the eye

18. Pecten: Fleshy comb with lots of blood vessels that sticks up in front of the retina—it likely provides the eye with nutrients and oxygen and removes waste

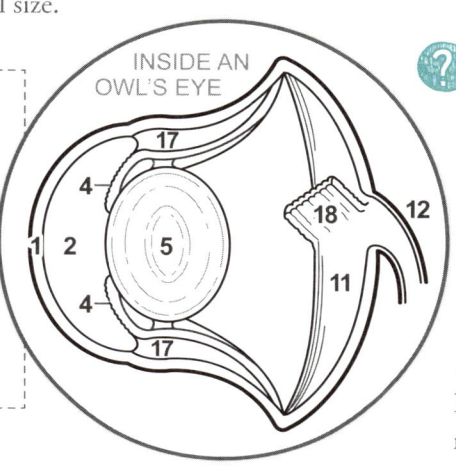

INSIDE AN OWL'S EYE

Did you know...

that birds do not have muscles around their eyeballs? They have to move their whole body or twist their necks from side to side to see.

Fun Fact!

Hippopotamuses have a thick, clear **nictitating membrane** (16) to protect their eyes and help them see under water. It's like having goggles!

CROCODILE

Fun Fact!

Crocodiles and **hippos** have protruding eyes on flat heads, so they can keep their eyes, ears, and nose out of the water at the same time.

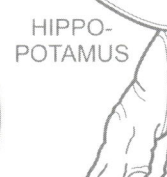

HIPPO-POTAMUS

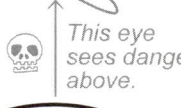

💀 *This eye sees danger above.*

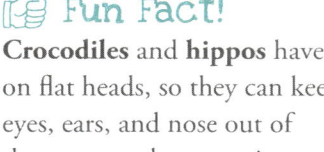

SHARK EYES

High light   Low light

Did you know...

that a **shark** cornea is so similar to our cornea that it can be transplanted into a human eye to help a person see better? A shark pupil can change size—it widens in low light conditions to capture all available light.

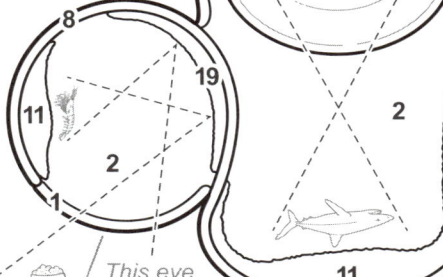

INSIDE A SPOOKFISH'S EYE

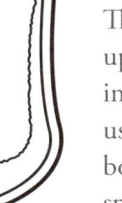

This eye sees food below.

Fun Fact!

Spookfish live in the deep oceans, and their eyes allow them to search for danger above and food below at the same time!

SPOOKFISH

HIPPO EYE

👁 Crazy Eyeballs!

The spookfish has 2 eyes on each side; 1 looks up, and 1 looks down. In the up-facing eye, images go through the lens to the retina as usual. But in the down-facing eye, images bounce off a **mirror** (19)—found only in a spookfish—and across the eye to the retina.

The Ear

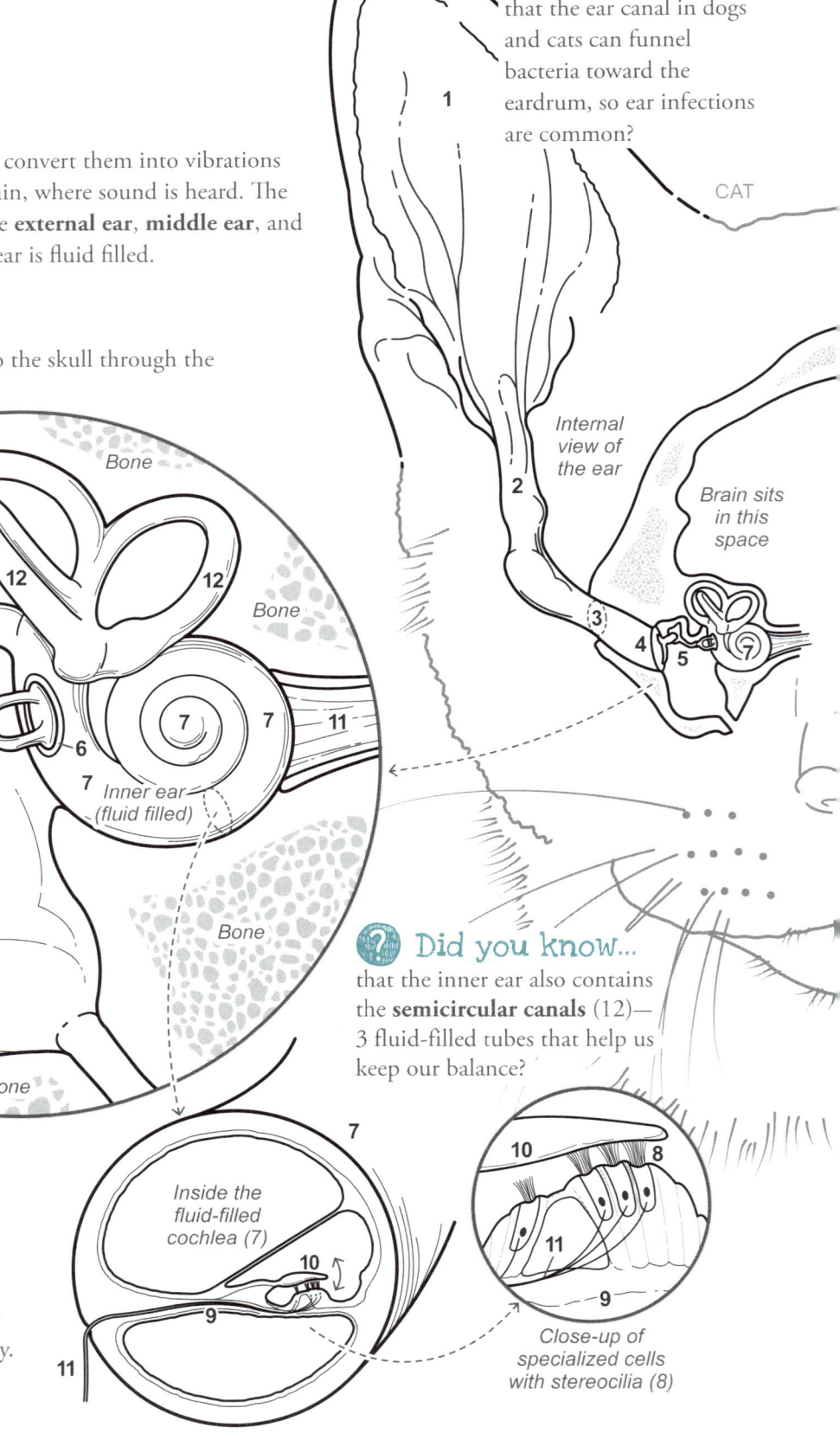

The purpose of the ear is to take sound waves from the air and first convert them into vibrations in a watery fluid and then into nerve impulses that travel to the brain, where sound is heard. The anatomy of the ear is quite complicated and has 3 main regions: the **external ear**, **middle ear**, and **inner ear**. The external and middle ear are air filled, but the inner ear is fluid filled.

External Ear
The fleshy **pinna** (1) and the **ear canal** (2) funnel sound waves into the skull through the **ear hole** (3) to the **eardrum** (4). The eardrum is a thin sheet (membrane) that vibrates when sound hits it.

Middle Ear
The eardrum vibrates **3 tiny bones** (5) in the middle ear: the hammer, anvil, and stirrup. These bones push on the **oval window** (6), a membrane separating the middle and inner ear. This converts air vibrations in the middle ear to fluid vibrations in the inner ear.

Inner Ear
Fluid vibrations travel inside the spirals of a snail-like tube called the **cochlea** (7). Special cells in the cochlea have little hairs called **stereocilia** (8). These are attached between a bony, **non-moving membrane** (9) and a **membrane that moves** (10) when the fluid vibrates. When this moveable membrane vibrates, the stereocilia stretch and contract, and this motion causes them to send impulses along the **nerves** (11) to the brain. This is when you hear a sound!

? Did you know...
that each of the cells in the cochlea responds to 1 sound frequency? As you get older, the high-frequency cells stop working as effectively.

? Did you know...
that the ear canal in dogs and cats can funnel bacteria toward the eardrum, so ear infections are common?

? Did you know...
that the inner ear also contains the **semicircular canals** (12)— 3 fluid-filled tubes that help us keep our balance?

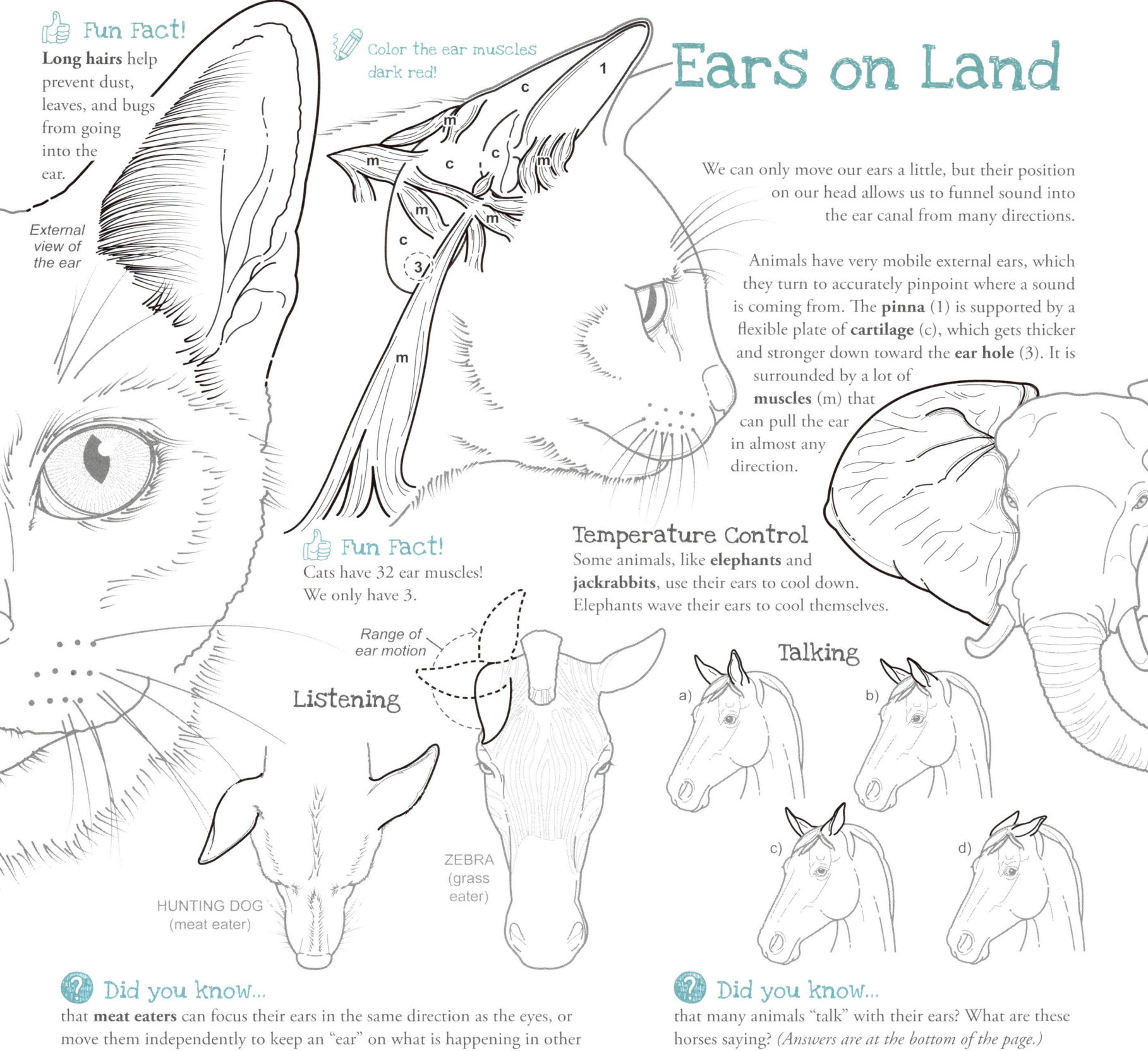

Ears in Air

Many flying animals hunt using sight, but owls also use extremely sensitive hearing.

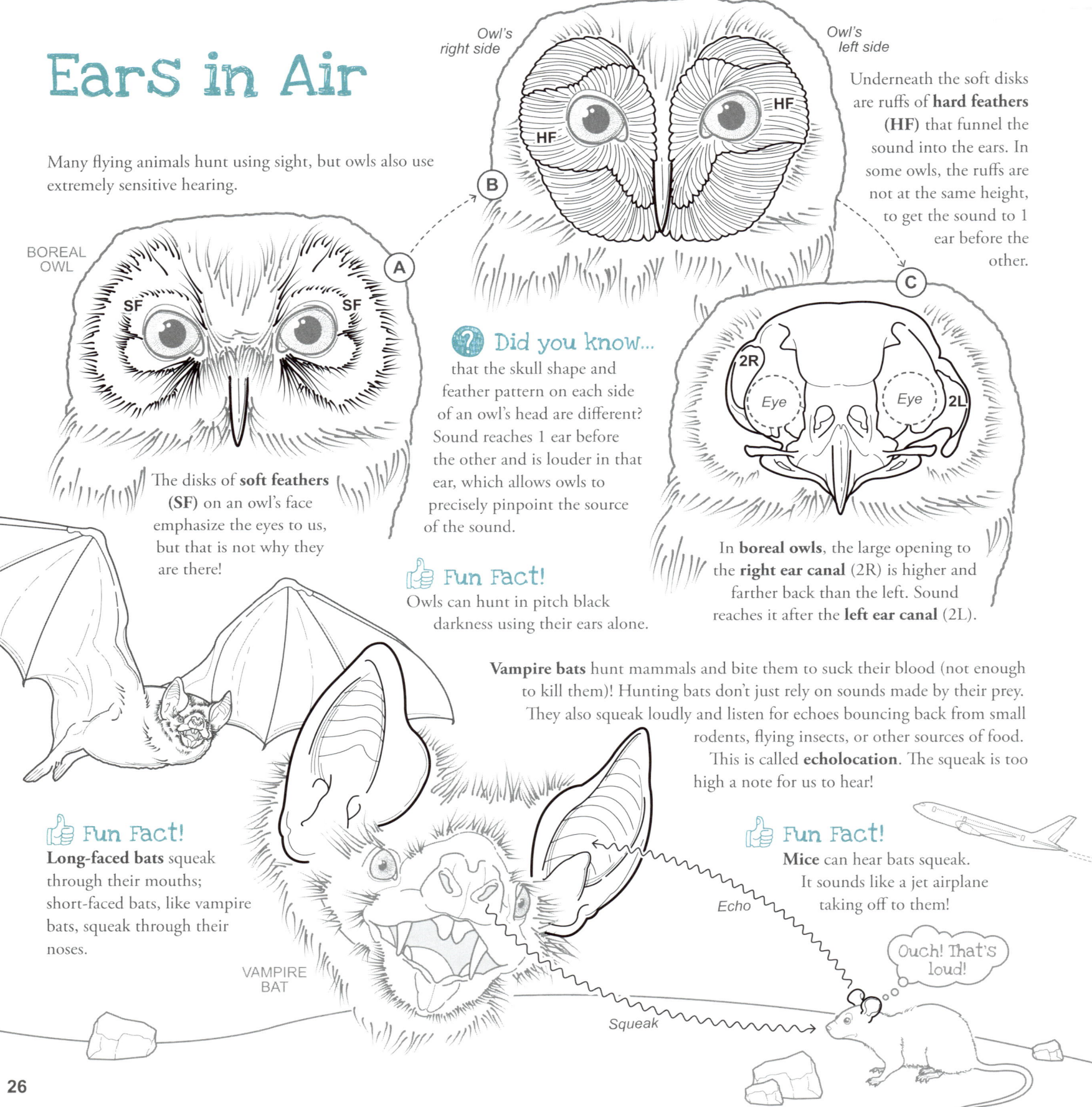

Underneath the soft disks are ruffs of **hard feathers (HF)** that funnel the sound into the ears. In some owls, the ruffs are not at the same height, to get the sound to 1 ear before the other.

The disks of **soft feathers (SF)** on an owl's face emphasize the eyes to us, but that is not why they are there!

🔎 Did you know...

that the skull shape and feather pattern on each side of an owl's head are different? Sound reaches 1 ear before the other and is louder in that ear, which allows owls to precisely pinpoint the source of the sound.

👍 Fun Fact!

Owls can hunt in pitch black darkness using their ears alone.

In **boreal owls**, the large opening to the **right ear canal** (2R) is higher and farther back than the left. Sound reaches it after the **left ear canal** (2L).

Vampire bats hunt mammals and bite them to suck their blood (not enough to kill them)! Hunting bats don't just rely on sounds made by their prey. They also squeak loudly and listen for echoes bouncing back from small rodents, flying insects, or other sources of food. This is called **echolocation**. The squeak is too high a note for us to hear!

👍 Fun Fact!

Long-faced bats squeak through their mouths; short-faced bats, like vampire bats, squeak through their noses.

👍 Fun Fact!

Mice can hear bats squeak. It sounds like a jet airplane taking off to them!

Ouch! That's loud!

Ears in Water

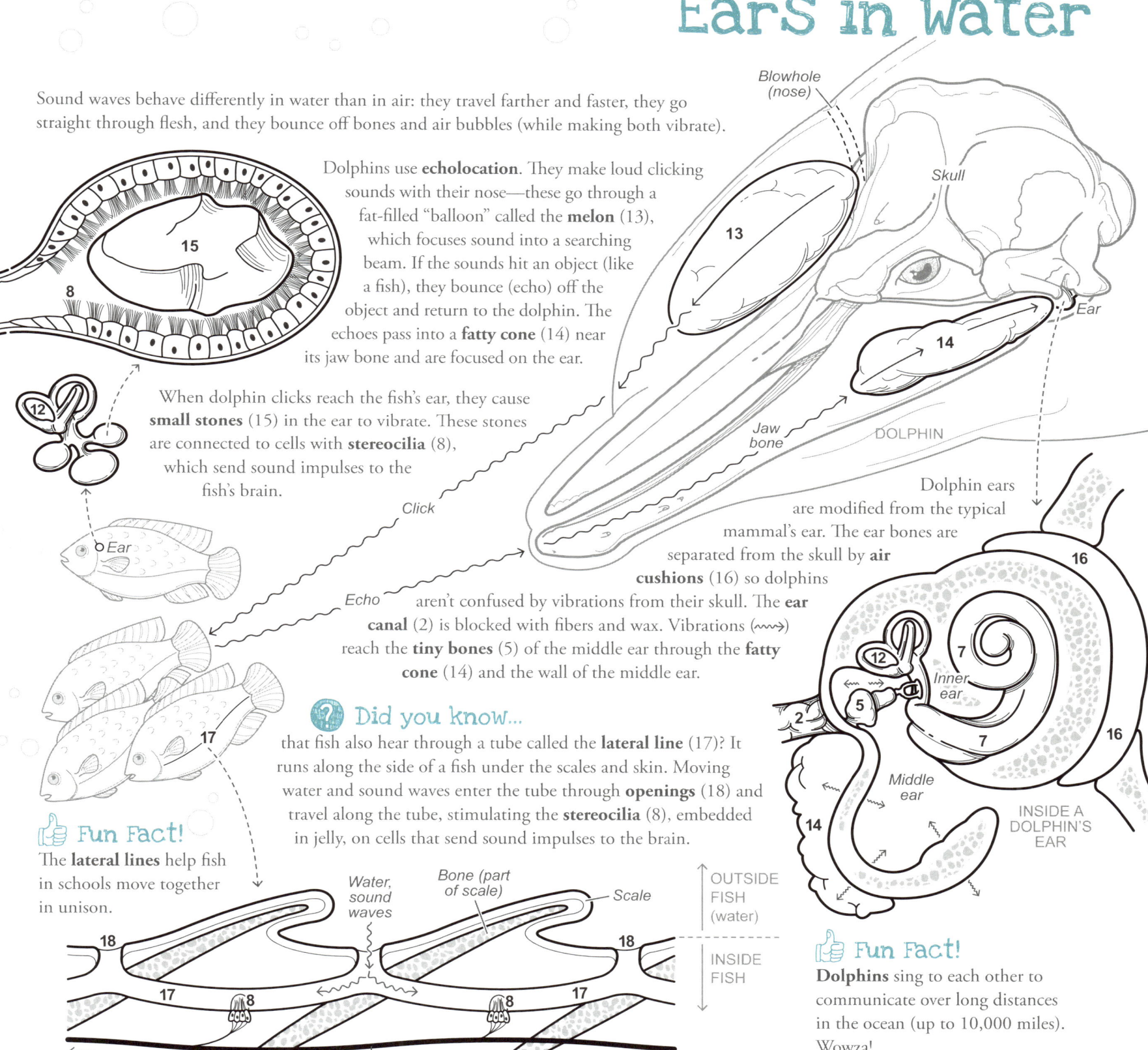

Sound waves behave differently in water than in air: they travel farther and faster, they go straight through flesh, and they bounce off bones and air bubbles (while making both vibrate).

Dolphins use **echolocation**. They make loud clicking sounds with their nose—these go through a fat-filled "balloon" called the **melon** (13), which focuses sound into a searching beam. If the sounds hit an object (like a fish), they bounce (echo) off the object and return to the dolphin. The echoes pass into a **fatty cone** (14) near its jaw bone and are focused on the ear.

When dolphin clicks reach the fish's ear, they cause **small stones** (15) in the ear to vibrate. These stones are connected to cells with **stereocilia** (8), which send sound impulses to the fish's brain.

Dolphin ears are modified from the typical mammal's ear. The ear bones are separated from the skull by **air cushions** (16) so dolphins aren't confused by vibrations from their skull. The **ear canal** (2) is blocked with fibers and wax. Vibrations (〰️) reach the **tiny bones** (5) of the middle ear through the **fatty cone** (14) and the wall of the middle ear.

❓ Did you know...

that fish also hear through a tube called the **lateral line** (17)? It runs along the side of a fish under the scales and skin. Moving water and sound waves enter the tube through **openings** (18) and travel along the tube, stimulating the **stereocilia** (8), embedded in jelly, on cells that send sound impulses to the brain.

👍 Fun Fact!
The **lateral lines** help fish in schools move together in unison.

👍 Fun Fact!
Dolphins sing to each other to communicate over long distances in the ocean (up to 10,000 miles). Wowza!

27

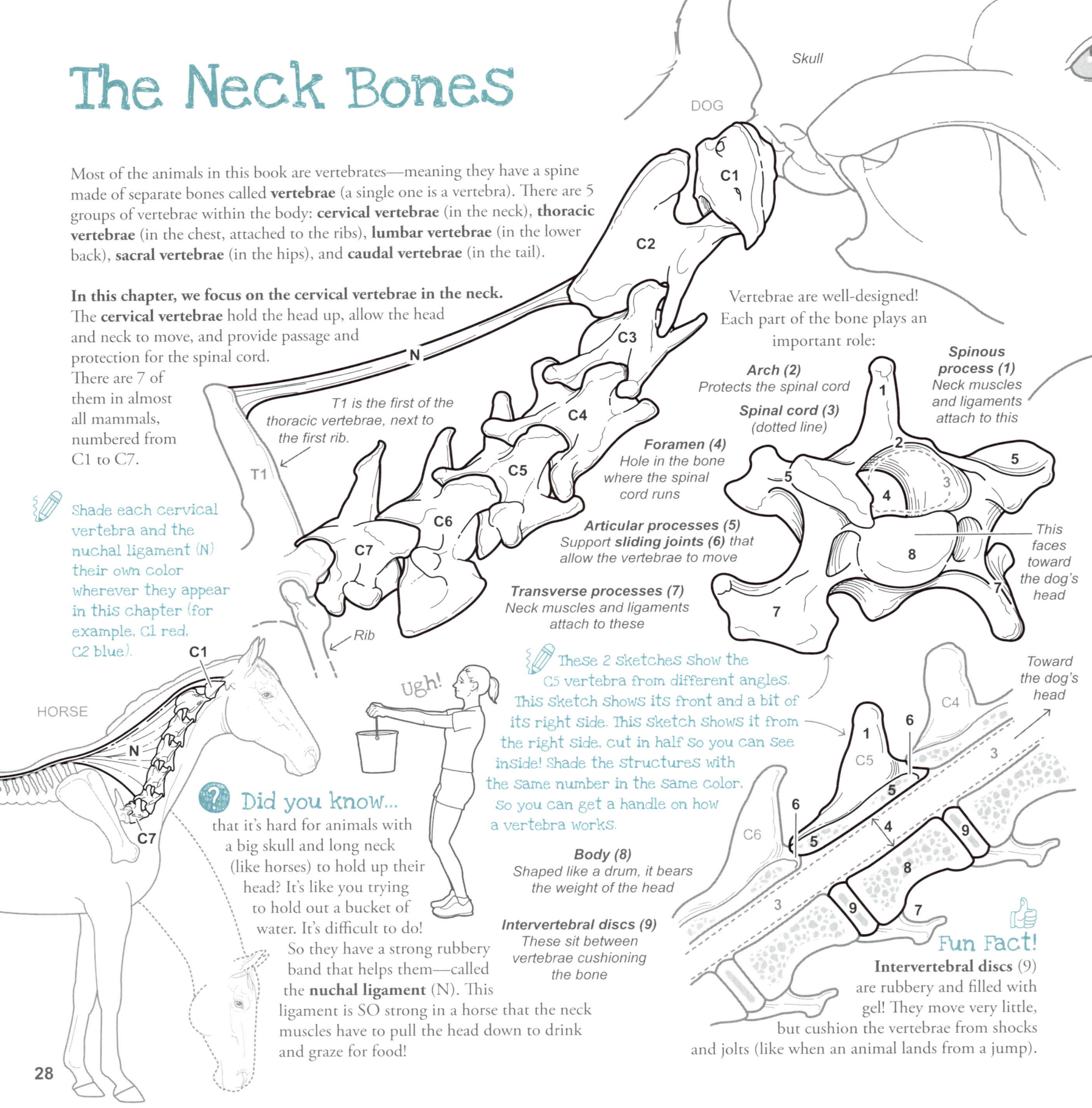

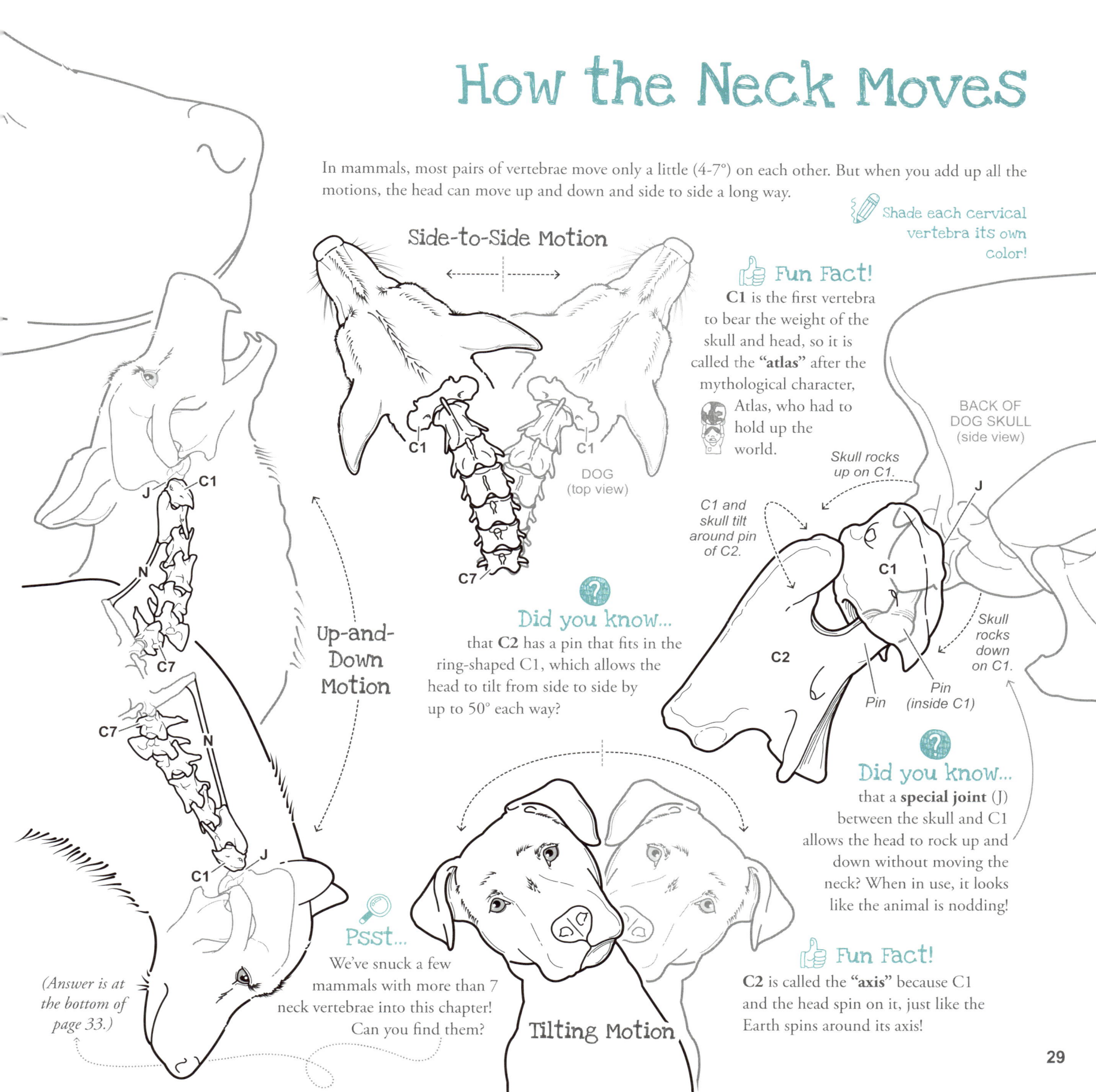

Long Necks

The longest necks in living animals are found in giraffes, but they still have **only 7 cervical vertebrae**! Each **vertebral body** (8) is much longer than ours—over a foot long in the largest males. A long neck helps the giraffe reach high into the trees to eat leaves and buds, pulling them off with a strong, snake-like tongue. The head rocks upward on the **atlas** (C1) to help it reach even higher.

Giraffes live on the savanna (grasslands) of Africa. On the other side of the Atlantic Ocean, in South America, lives another leaf-eating animal, the **3-toed sloth**. Hmm... can you guess why we're showing you a sloth here? Look closely...

? Did you know...
that, like a horse, a giraffe has a long **nuchal ligament** (N) anchored at its shoulders to help hold up its head and long neck?

? Did you know...
that giraffe necks are very flexible, despite only having 7 vertebrae? The ends of the vertebrae form **cup-and-ball joints**, which allow the neck to be strong but supple.

? Did you know...
that a giraffe can rock the **first vertebra in its chest** (T1)? This rocking motion rotates the whole neck a little bit more, enabling the giraffe to reach further downward.

👍 Fun Fact!
Giraffe necks are built for reaching up, not down. So when a giraffe wants to drink, its legs have to help the head reach the water by spreading far to the sides.

👍 Fun Fact!
The longest necks ever are found in some dinosaurs.

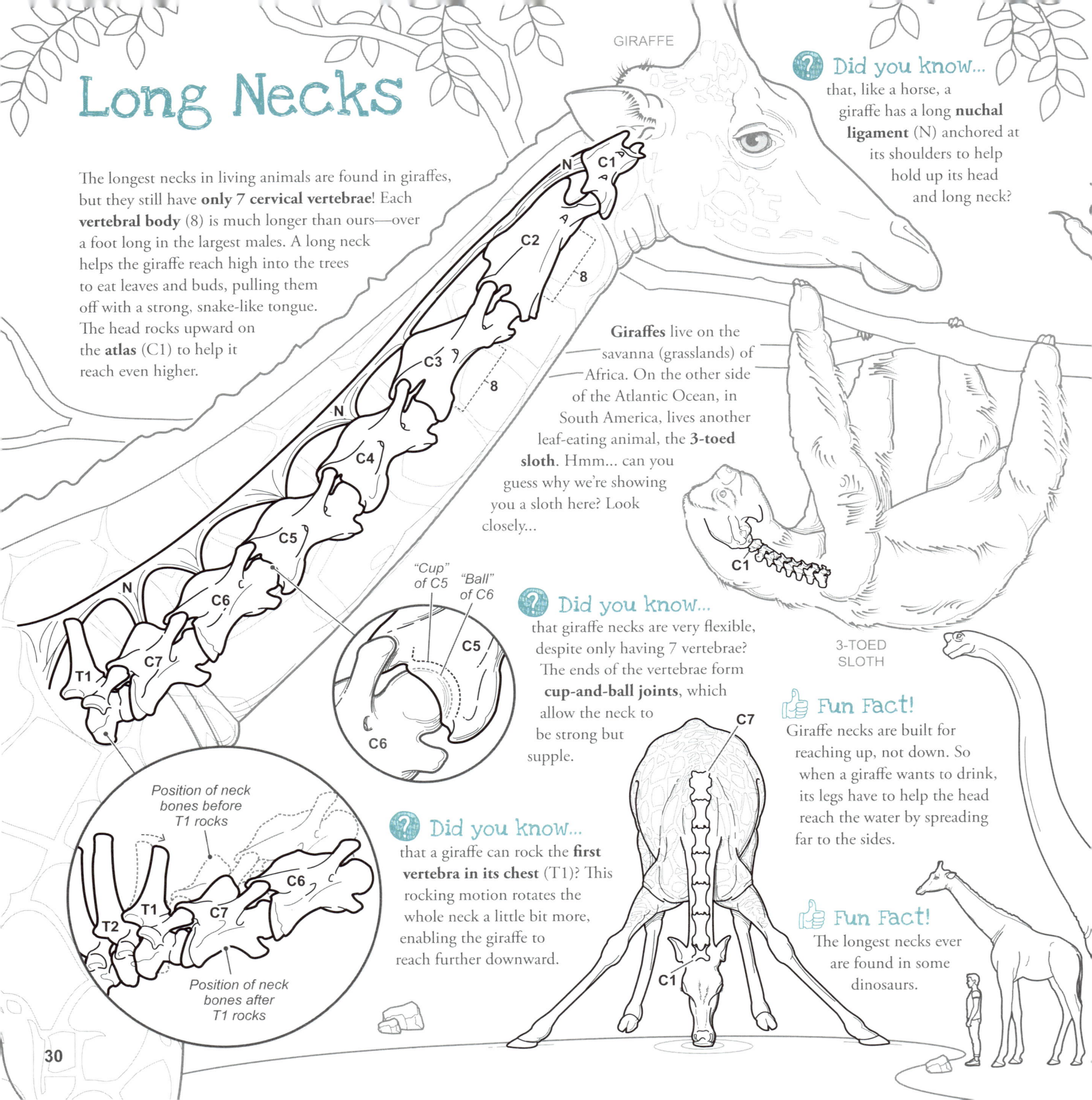

GIRAFFE

3-TOED SLOTH

"Cup" of C5
"Ball" of C6

Position of neck bones before T1 rocks
Position of neck bones after T1 rocks

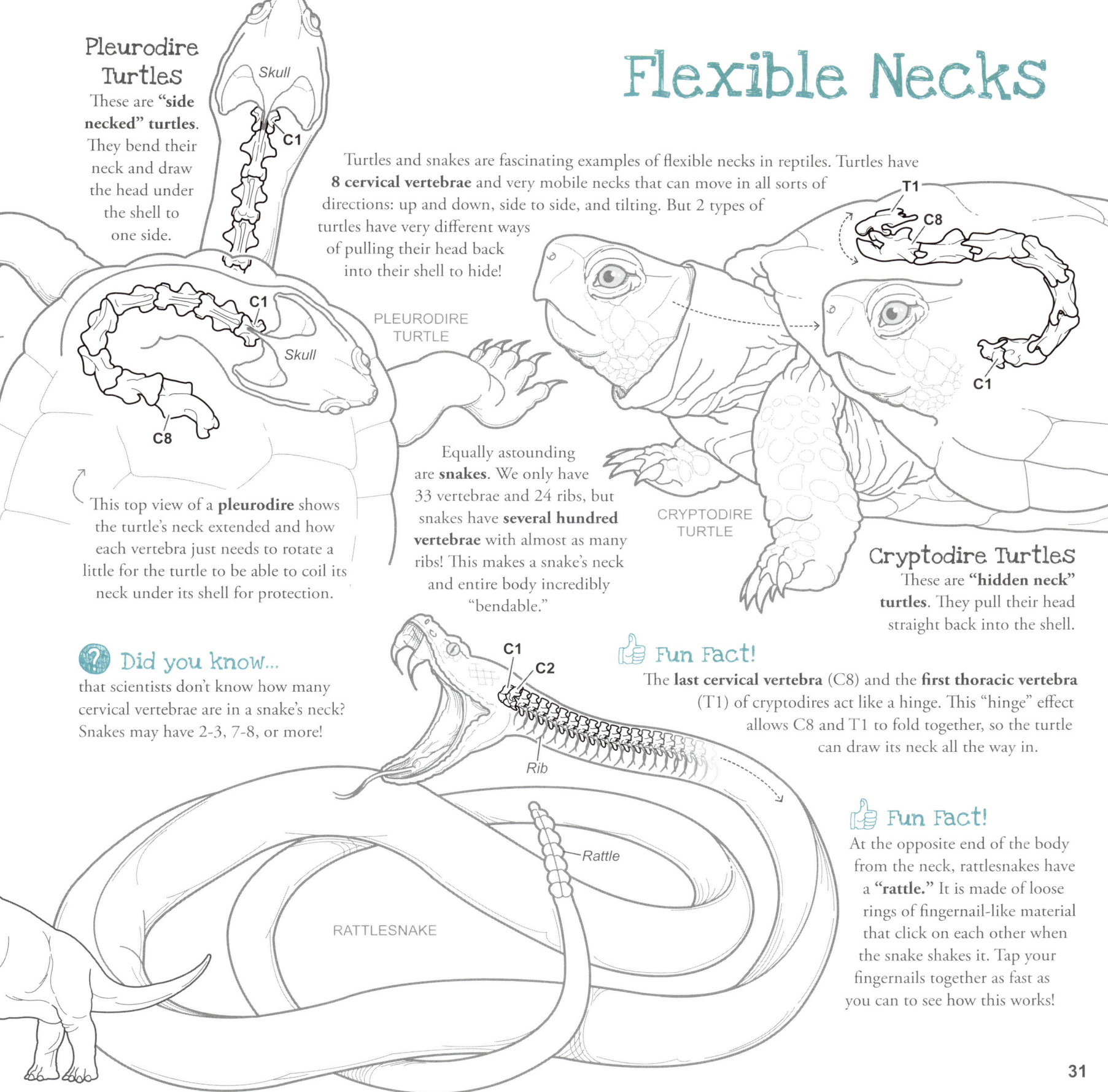

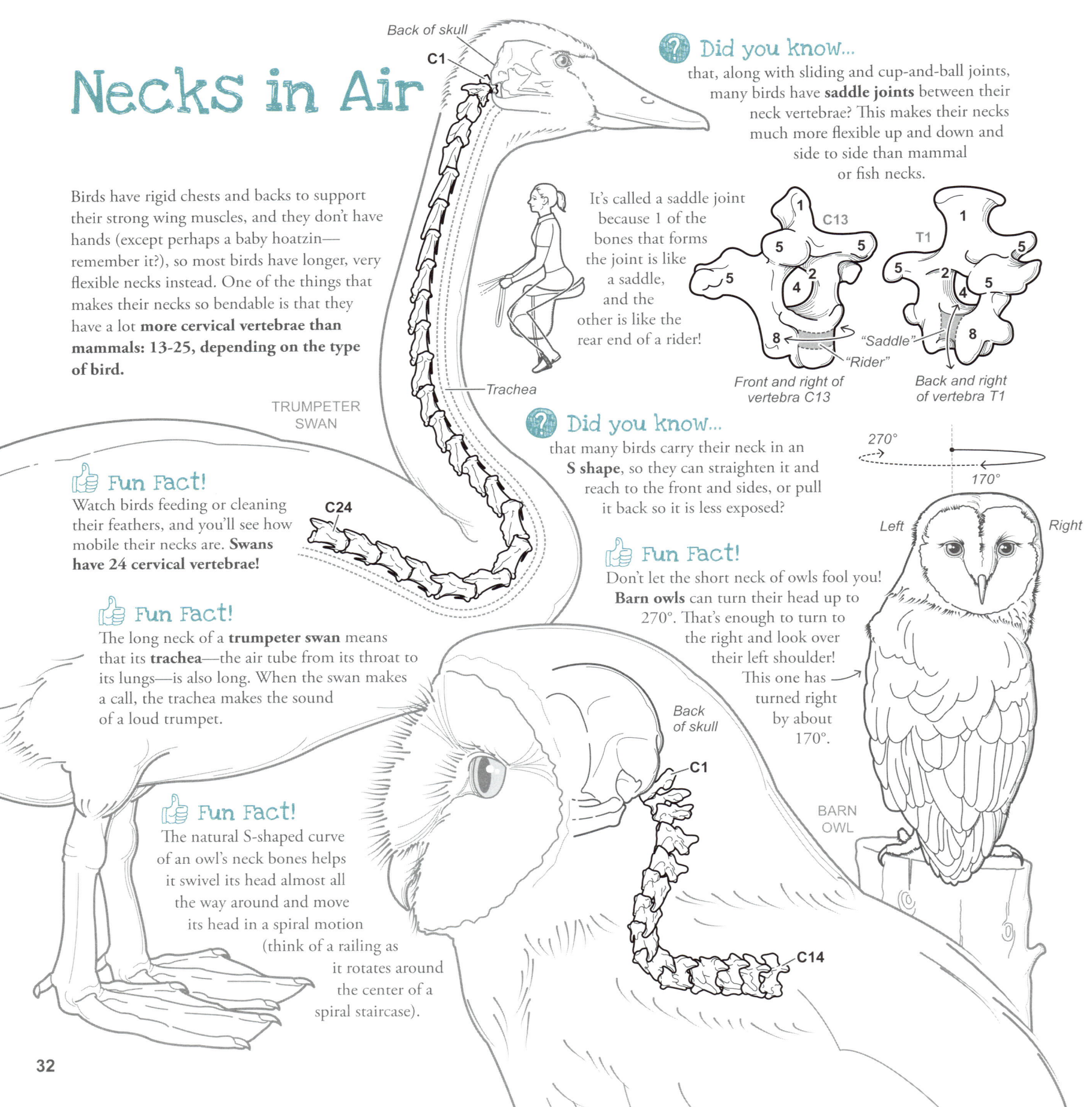

Necks in Water

Holding their bodies (and necks) up against the force of gravity is less of an issue for animals living in water because water is buoyant, so their cervical vertebrae (in fact, all their vertebrae) have a simpler structure than in animals living on land.

In fast-swimming mammals—**dolphins** and **whales** (yep, whales are surprisingly fast!)—holding their head steady against the pushback of water would be difficult with a long, flexible neck. So their neck is shortened and stiffened by flattening and fusing together some or all of the cervical vertebrae.

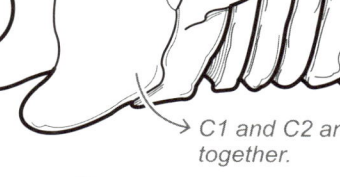

C1 and C2 are fused together.

DOLPHIN

Strange but true...

Compare the necks of **dolphins** and **giraffes**. They have very different body shapes with very different sized necks, living in very different environments—but both have the exact same number of cervical vertebrae! Isn't that amazing?

Slow-swimming, gentle **manatees** are common in the coastal waters of the eastern US. Notice that they have longer vertebrae than dolphins—this allows them to bend their necks while feeding. Hmm... what else do you notice? *(Hint: Count them!)*

👍 Fun Fact!

A light skeleton helps prevent fish from sinking. **Dolphins** are able to have a heavier skeleton than fish because they have lungs filled with air, which keeps them buoyant.

🐟 By the way...

Fish don't have cervical vertebrae! Their skull is connected directly to the thoracic vertebrae. It is also connected to the pectoral (and sometimes the pelvic) fins by a few small bones.

Are You a Genius?

We talked about these bones on page 12. Do you remember their names? Consider yourself a genius if you do!
(Answer is at the bottom of the page.)

Pectoral fin

Pelvic fin

PERCH

Back of skull

MANATEE

Mammals that don't have 7 cervical vertebrae: 3-toed sloth (8-9), 2-toed sloth (5-7), and manatee (6). Birds have more than 7, but they aren't mammals! **Are You a Genius? answers:** Clavicle, scapula, and coracoid.

The Heart

Heart Parts
- 1a. Right atrium
- 1b. Left atrium
- 2a. Right ventricle
- 2b. Left ventricle
- 3. Chordae tendineae
- 4. Papillary muscle
- 5. Trabeculae
- 6. Interventricular septum
- 7. Atrioventricular (AV) valves
- 8. Semilunar valves
- 9. Cranial vena cava
- 10. Caudal vena cava
- 11. Pulmonary arteries
- 12. Pulmonary veins
- 13. Aorta
- 14. Coronary arteries

All mammal hearts have 4 chambers, from a huge blue whale to the tiniest shrew. The right and left sides of the heart are muscular pumps—each side has a thin-walled collecting chamber called an **atrium** (1) that sits atop and fills a larger, thick-walled pumping chamber called a **ventricle** (2). These pumps keep blood flowing through the body, and the oxygen and nutrients contained in blood keep body cells alive and active.

Blood Flow within the Heart
Blood returns to the heart from the body in tubes called veins. The blood is dark red and low in oxygen (body has used up all the oxygen). Blood enters the **right atrium** (1a) through the largest veins—the **venae cavae** (9, 10)—and is pumped to the lungs, where oxygen is added, turning it bright red. This high-oxygen blood is sent to the left side of the heart, where the **left ventricle** (2b) pumps it out to the rest of the body in tubes called arteries—the **aorta** (13) is the first and largest of these.

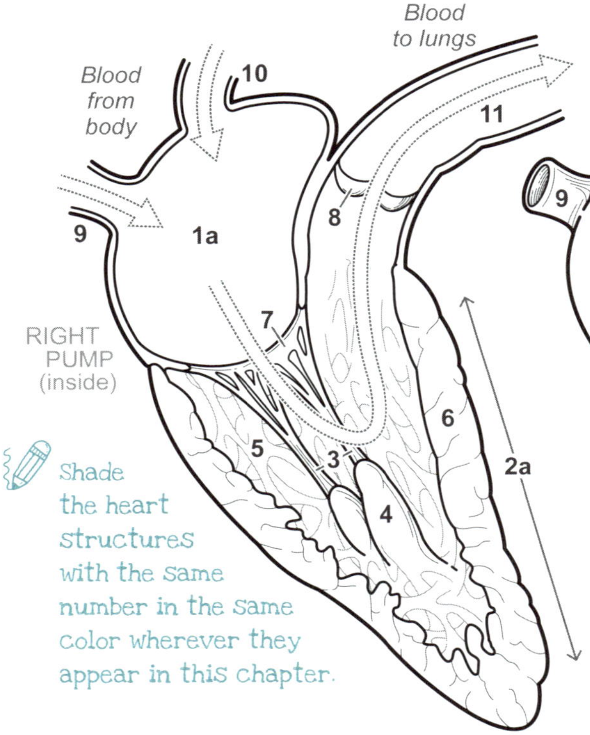

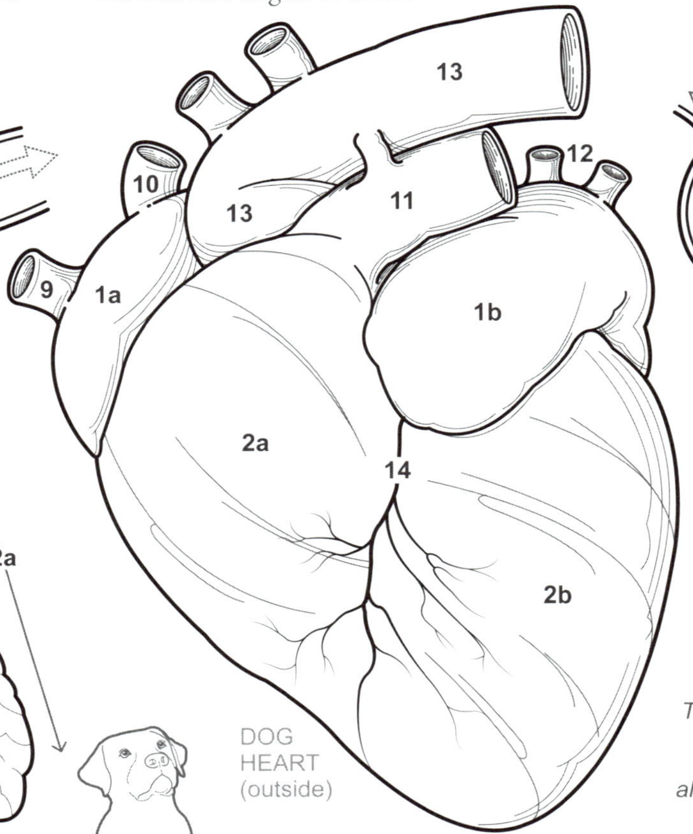

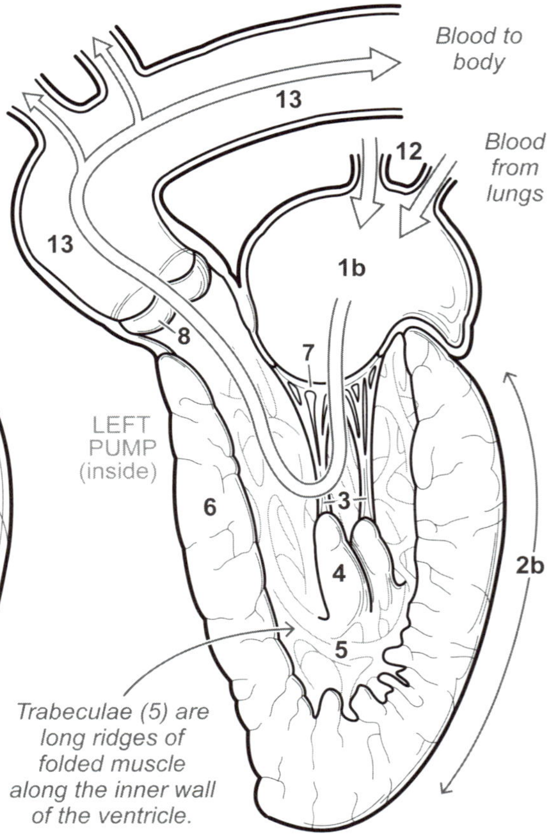

Shade the heart structures with the same number in the same color wherever they appear in this chapter.

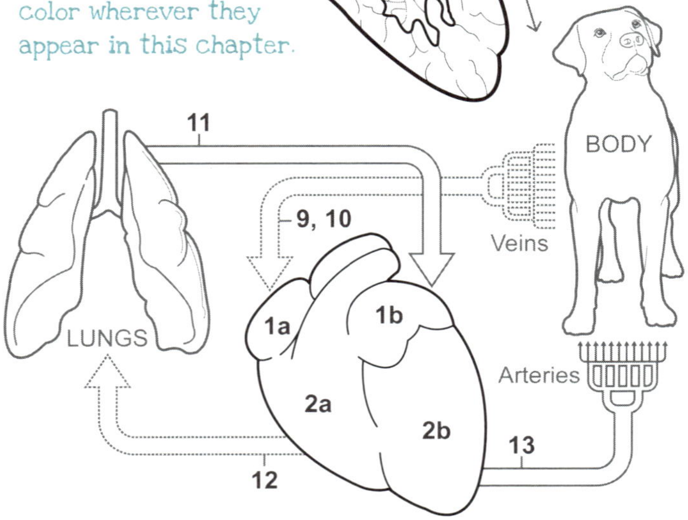

Trabeculae (5) are long ridges of folded muscle along the inner wall of the ventricle.

Blood To & From the Heart
Big arteries are like highways—and they divide into smaller and smaller roads. The smallest blood "roads" are **capillaries**. These run between all the cells of the body delivering oxygen and nutrients. They also connect the smallest arteries to the smallest veins. As veins progress back to the heart, they merge into larger and larger veins —from 1-lane tracks to major highways.

♥ Hmmmm...
Did you notice that the left ventricle wall is much thicker than the right ventricle wall? Can you think why? (*Look at answer (a) at the bottom of page 39.*)

The Heart

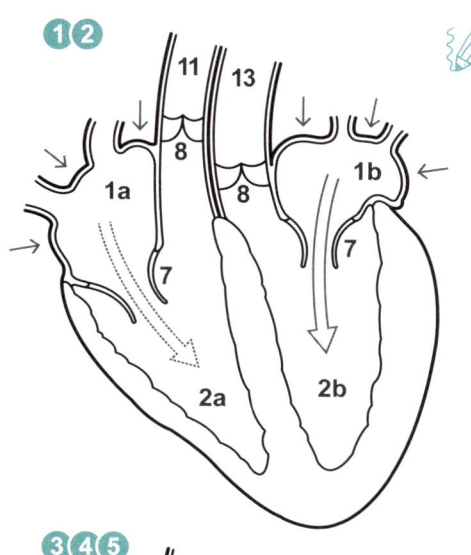

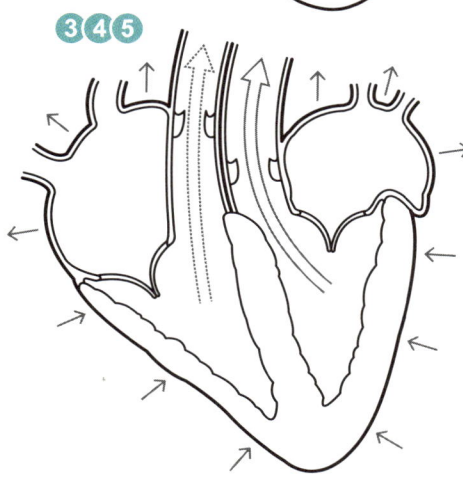

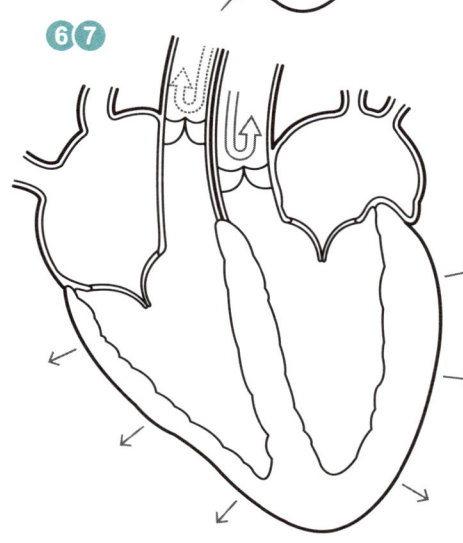

✏️ Color high-oxygen blood flow (⇨) RED and low-oxygen blood flow (⇨) BLUE wherever they appear in this chapter.

How the Heart Works

The heart is continually relaxing (to fill with blood) and contracting (to pump blood out). Gently put your ear on your pet's chest, and listen carefully. **Each heartbeat has 2 sounds**—a loud LUB and a softer DUP. Here's how it happens...

1. The atria (1) contract to push blood into the relaxed ventricles (2).
2. This causes the AV valves (7) between each atrium and ventricle to open wide. Blood flows into the ventricles.
3. The atria relax, and the 2 ventricles contract.
4. Blood pushes against the semilunar valves (8), forcing them open, while also forcing the AV valves to close. The LUB sound is made when the AV valves snap shut.
5. Blood flows into the pulmonary artery (11) and aorta (13) on its way to the lungs and body.
6. As the entire heart relaxes (briefly!), the semilunar valves close tight, preventing blood in the arteries from rushing back into the ventricles. As they shut, they make the DUP sound.
7. Then the whole cycle repeats itself!

❓ Did you know...

that if high- and low-oxygen blood mixed, an animal would feel tired all the time? It can't supply enough oxygen for the body, which is why it doesn't mix in the heart! Arteries only carry **high-oxygen blood FROM the heart**, while veins only carry **low-oxygen blood TO the heart**. Can you find the 2 exceptions to this rule? *(Look at answer (b) at the bottom of page 39.)*

👍 Fun Fact!

The heart rate or **beats per minute (bpm)** of different animals varies wildly...

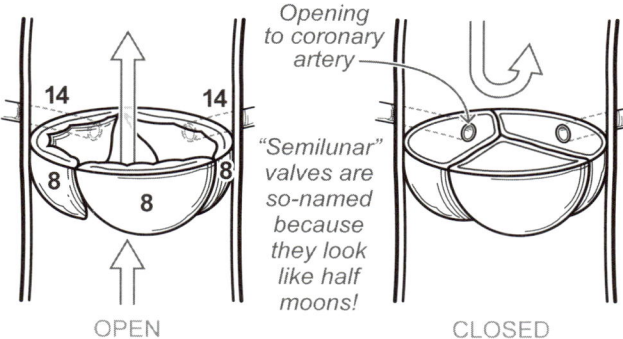

OPEN   CLOSED

Opening to coronary artery

"Semilunar" valves are so-named because they look like half moons!

❓ Did you know...

that **semilunar valves** (8) are like 3 jean back-pockets? When blood tries to backflow into the ventricle, the pockets fill with blood, pushing their edges together. This shuts the ventricle, but allows blood to flow into the **coronary arteries** (14), which supply the heart muscles with blood.

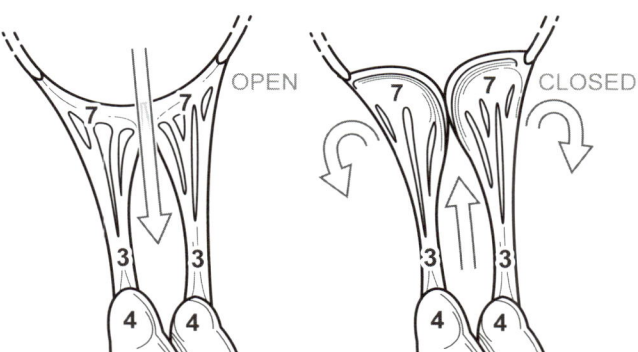

OPEN   CLOSED

❓ Did you know...

that **AV valves** (7) are like parachutes? Made of thin, flexible tissue, they're held in position with strong elastic strings—**chordae tendineae** (3)—which are anchored to **muscles** (4) in the ventricle wall.

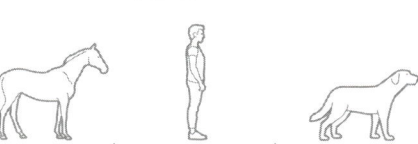

WHALE	HORSE	HUMAN	DOG	CHEETAH	HAMSTER	PYGMY SHREW
20 bpm	32 bpm	60 bpm	80 bpm	120 bpm	450 bpm	1,500 bpm

 These are the heart rates of resting animals. When running, horse and cheetah hearts can speed up to 280 bpm. Imagine the rate for a sprinting pygmy shrew! Wowza!

3-Chamber Hearts

Unlike mammals, **most reptiles** (like lizards, turtles, and snakes) and **all amphibians** (like frogs, toads, salamanders, and newts) have **3-chambered hearts**.

Let's explore the 3-chambered heart of a Komodo dragon (a huge lizard)! Find the **sinus venosus** (15)—the enlarged area just before the atrium (in mammals, it is merged with the right atrium and not a separate part). It empties low-oxygen blood collected from the body into the **right atrium** (1a). Now find the **left atrium** (1b), which fills with high-oxygen blood from the lungs. Trace the blood as it flows from both atria into 1 **ventricle** (2) and out to the lungs through the **pulmonary artery** (11) and to the body through the **aortae** (13). But, wait! Only 1 ventricle means the low- and high-oxygen blood are going to mix, right? No need to worry—the ventricle is packed with **trabeculae** (5), making it look like a sponge, and this controls the amount of mixing!

Color high-oxygen blood flow ⟹ RED and low-oxygen blood flow ⇢ BLUE!

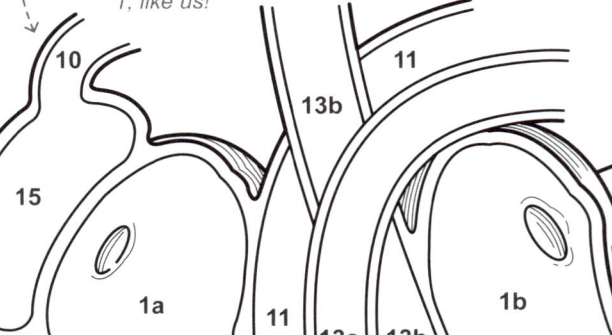

KOMODO DRAGON

Notice that reptiles have 2 aortae (13a, b), not just 1, like us!

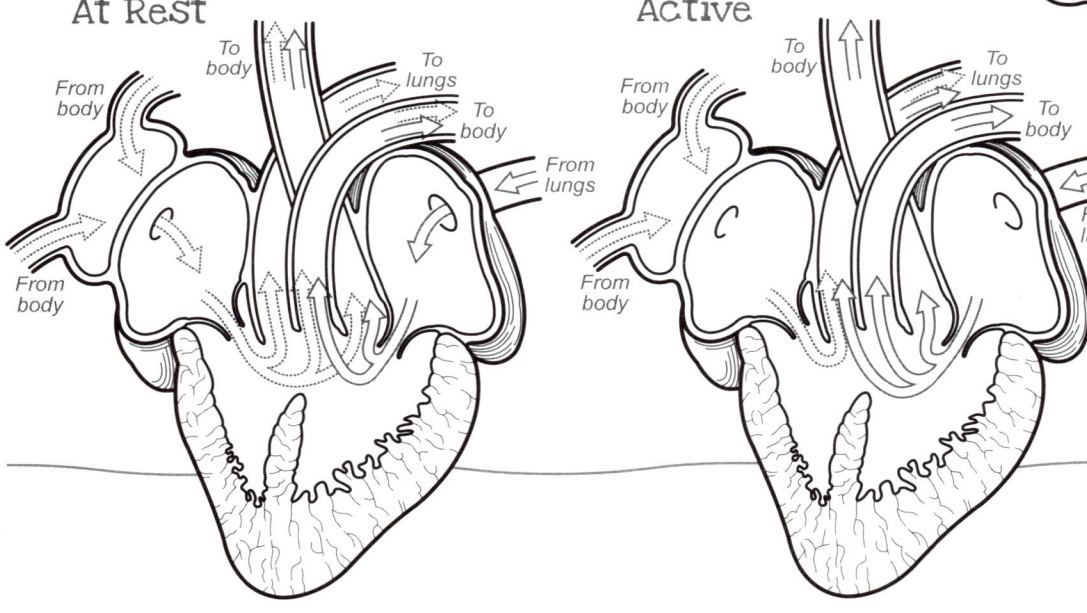

INSIDE A KOMODO DRAGON'S HEART

When a Komodo dragon is resting, the trabeculae allow low-oxygen blood to spread from the right side of the ventricle to the left, which mixes a bit with the high-oxygen blood going into the aortae. This is not a problem because the body doesn't need as much oxygen when at rest.

When a Komodo dragon is active, the trabeculae on the left side of the ventricle contract more forcefully, pushing high-oxygen blood from the left side of the ventricle to the right and into both aortae, so the working body gets lots of oxygen.

❓ Did you know...

that the **ventricles** (2a, b) of alligators are almost completely divided by a **septum** (6) like mammals, but have a connection at the top called the **foramen of Panizza** (16)?

ALLIGATOR

2-Chamber Hearts

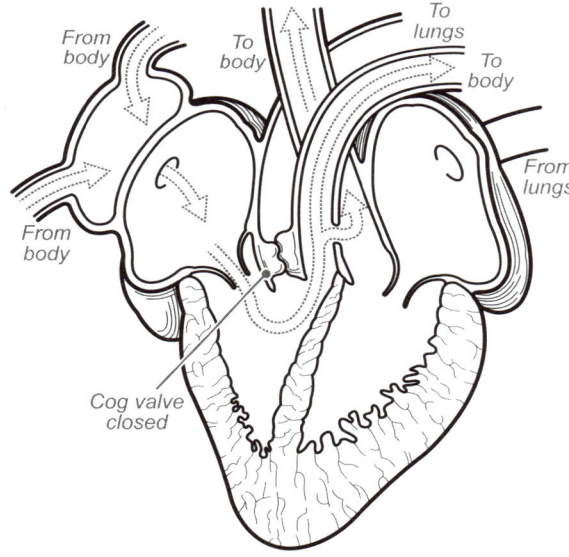

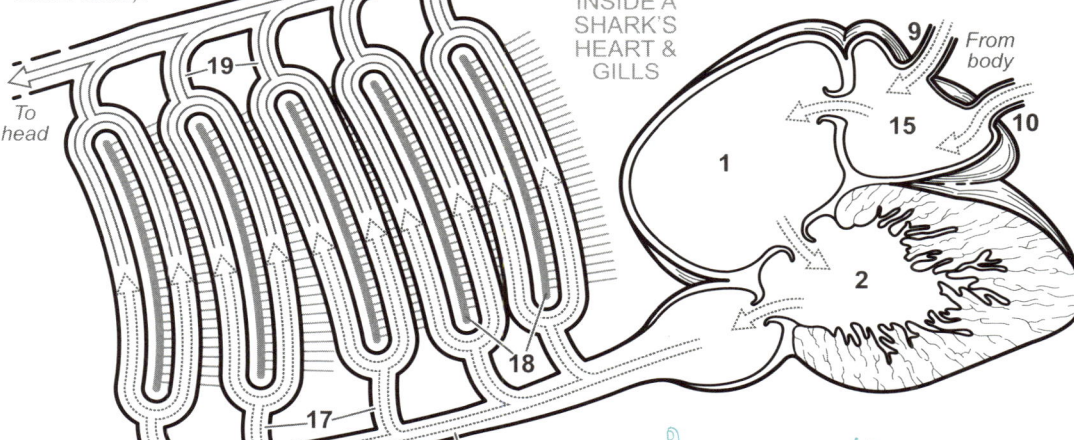

Fish have a very simple heart, with only 2 chambers—1 atrium and 1 ventricle. They have gills instead of lungs, so instead of having 2 pumps like mammal hearts—1 to push blood to the lungs and the other to push blood to the body—fish just need 1 pump to push blood past the gills and to the rest of the body. Like mammals, fish keep high- and low-oxygen blood separate, but in a very different way.

👍 Fun Fact!

The heart rate of a slowly swimming, medium-sized **shark** is 20-50 bpm. Compare this to the heart rate of a dog, which is about 60-120 bpm depending on the dog's size (larger dogs have lower heart rates).

When alligators are breathing air, hardly any blood flows through the **foramen of Panizza** (16). But when they dive under water, the blood flow to and from the lungs is shut off using a unique valve—called a **cog valve**—and more blood flows through the foramen. This helps the heart recirculate blood to and from the body. With no oxygen being added from the lungs, the oxygen in the blood continues to drop until the alligator surfaces for air.

Here's how a shark heart works. Low-oxygen blood—from the body and traveling in the **venae cavae** (9, 10)—enters the **sinus venosus** (15), which empties into the **single atrium** (1). It flows into the **ventricle** (2) and is then pumped into the **lower aorta** (13a), which directs the low-oxygen blood into **arteries** (17) leading to the **gills** (18). The gills have many capillaries that collect oxygen from the water and transfer it to the passing blood. This high-oxygen blood is collected in **arteries** (19) leading away from the gills and into the **upper aorta** (13b), which branches into arteries that take the high-oxygen blood to all parts of the body.

👍 Fun Fact!

Sharks don't need to pump blood as fast as dogs, because their cells use much less oxygen and energy. Sharks are **"cold-blooded"**—their body temperature is similar to the cold water they live in—so they don't need to use up energy to keep warm—a major energy-saver!

37

Unusual Hearts

Octopuses have 3 hearts! The **vena cava** (9) brings low-oxygen blood from the body to the 2 **side hearts** (20). They pump blood through the **gills** (18), where it picks up oxygen on the way to the **central heart** (21). The central heart pumps the high-oxygen blood (and nutrients) to the body through the **aorta** (13).

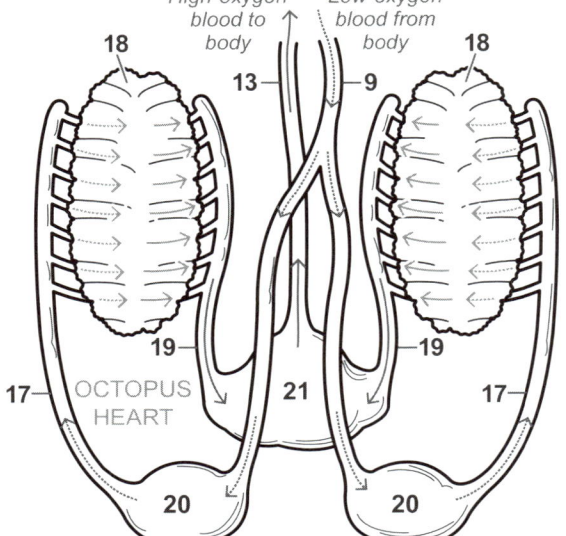

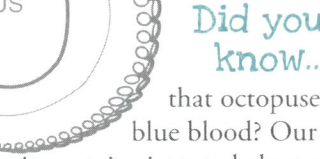

OCTOPUS

👍 Fun Fact!

The **central heart** (21) stops beating when an octopus swims. They tire very quickly without the flow of blood and oxygen—which is why octopuses prefer to crawl on the sea floor rather than swim.

❓ Did you know...

that when octopuses die, their blood becomes colorless? This is because it has lost all its oxygen.

❓ Did you know...

that octopuses have blue blood? Our blood is red because it contains iron to help carry oxygen to our body cells. But octopus blood uses copper to carry oxygen instead, turning their blood blue. This is ideal because copper is better at carrying oxygen in the cold and low-oxygen ocean where octopuses live.

COCKROACH

Cockroaches don't have veins, so tiny holes in the chambers allow used blood to get back into the heart.

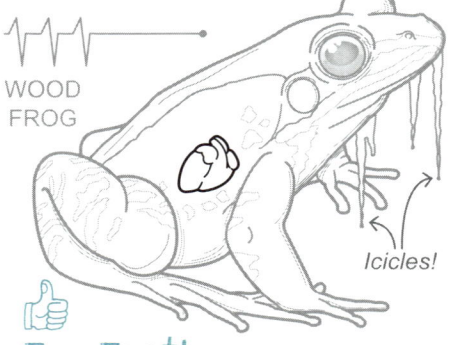

WOOD FROG

Icicles!

❓ Did you know...

that scientists don't call it insect "blood"? Instead, they call it **hemolymph**. This is because it is different than regular blood—it doesn't contain red blood cells, it doesn't carry oxygen, and it isn't red.

👍 Fun Fact!

A wood frog's heart stops beating and the frog freezes almost completely solid in the winter. It can stay like this for months at a time! As cold weather approaches, its body makes an antifreeze that protects its cells and keeps the frog alive until its blood flows again. When spring comes, wood frogs thaw, and their heart begins to beat.

The cockroach heart is very different from other hearts we've explored. It has 13 chambers, it doesn't beat by itself, and it doesn't help to pump oxygen to the body. The heart and **aorta** (13) form a long, tube-like structure running along the insect's back. **Muscles outside the heart** (22) are responsible for its beating. Blood enters the heart when these muscles relax and is pushed through the heart when they contract. Blood in the heart only moves in 1 direction: from the rear of the cockroach toward its head. Once at the head, the blood leaves the aorta and is not confined in arteries and veins. Instead, it flows freely around the insect's tissues and organs, until it is picked up by the heart and pumped forward once again.

Self-Healing Hearts

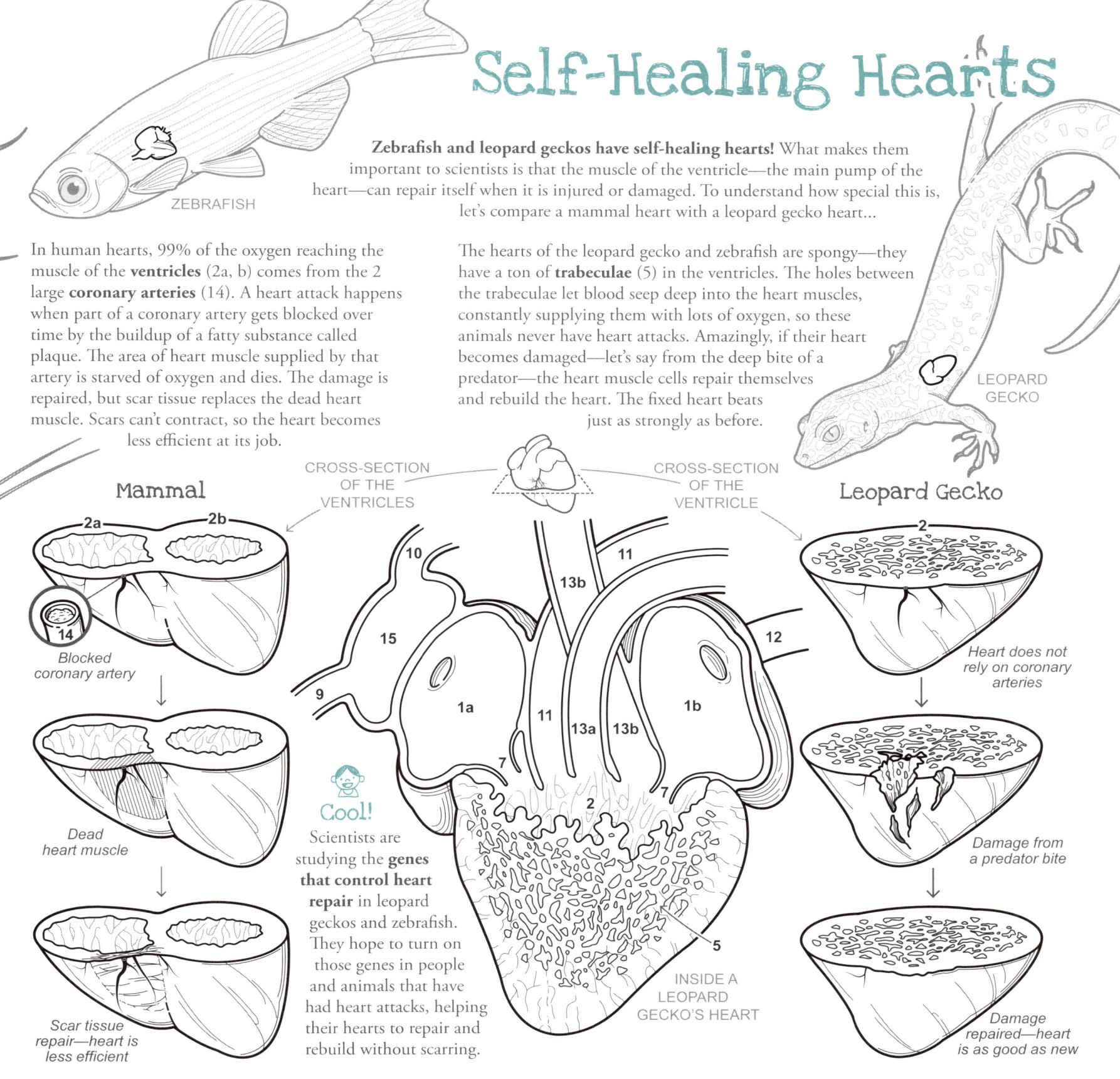

Zebrafish and leopard geckos have self-healing hearts! What makes them important to scientists is that the muscle of the ventricle—the main pump of the heart—can repair itself when it is injured or damaged. To understand how special this is, let's compare a mammal heart with a leopard gecko heart...

In human hearts, 99% of the oxygen reaching the muscle of the **ventricles** (2a, b) comes from the 2 large **coronary arteries** (14). A heart attack happens when part of a coronary artery gets blocked over time by the buildup of a fatty substance called plaque. The area of heart muscle supplied by that artery is starved of oxygen and dies. The damage is repaired, but scar tissue replaces the dead heart muscle. Scars can't contract, so the heart becomes less efficient at its job.

The hearts of the leopard gecko and zebrafish are spongy—they have a ton of **trabeculae** (5) in the ventricles. The holes between the trabeculae let blood seep deep into the heart muscles, constantly supplying them with lots of oxygen, so these animals never have heart attacks. Amazingly, if their heart becomes damaged—let's say from the deep bite of a predator—the heart muscle cells repair themselves and rebuild the heart. The fixed heart beats just as strongly as before.

ZEBRAFISH

LEOPARD GECKO

Mammal

- Blocked coronary artery
- Dead heart muscle
- Scar tissue repair—heart is less efficient

CROSS-SECTION OF THE VENTRICLES

CROSS-SECTION OF THE VENTRICLE

Leopard Gecko

- Heart does not rely on coronary arteries
- Damage from a predator bite
- Damage repaired—heart is as good as new

INSIDE A LEOPARD GECKO'S HEART

Cool! Scientists are studying the **genes that control heart repair** in leopard geckos and zebrafish. They hope to turn on those genes in people and animals that have had heart attacks, helping their hearts to repair and rebuild without scarring.

Answers: (a) Because the left ventricle has to pump blood to the entire body, while the right only pumps blood to the nearby lungs. (b) Pulmonary arteries (carry low-oxygen blood from the heart) and pulmonary veins (carry high-oxygen blood to the heart).

Illustration References/Credits

Front cover, page 23: Owl eye anatomy
This illustration was adapted from Figure 1 in:
Martin G. What drives bird vision? Bill control and predator detection overshadow flight. *Front Neurosci.* 2017;1:619. doi:10.3389/fnins.2017.00619

Pages 12, 33: Fish forelimb
Galloway TW. Pectoral girdle and fin of a teleost [illustration]. ClipArt ETC, Florida Center for Instructional Technology, College of Education, University of South Florida. Accessed December 17, 2018. https://etc.usf.edu/clipart/24900/24968/teleost_24968.htm

Thomson JA. Pectoral girdle and fin of cod [illustration]. ClipArt ETC, Florida Center for Instructional Technology, College of Education, University of South Florida. Accessed December 17, 2018. https://etc.usf.edu/clipart/48000/48052/48052_cod_fin.htm

Page 17: Cow gastrointestinal tract
The intestine in this illustration was in part adapted from Figure 1 in:
Reti KL, Thomas MC, Yanke LJ, Selinger LB, Inglis GD. Effect of antimicrobial growth promoter administration on the intestinal microbiota of beef cattle. *Gut Pathog.* 2013;5:8. doi:10.1186/1757-4749-5-8

Page 19: Shark gastrointestinal tract
The intestine in this illustration was in part adapted from Figure 3.25 in:
De Iuliis G, Pulerà D. The shark. In: De Iuliis G, Pulerà D, eds. *The Dissection of Vertebrates.* 2nd ed. Academic Press; 2011:27-77.

Page 20: Cellular layers of the retina
A stock image was adapted to create this illustration:
Designua. Eye anatomy [digital illustration]. Shutterstock.

Page 23: Spookfish eye
Partridge JC, Douglas RH, Marshall NJ, Chung W-S, Jordan TM, Wagner H-J. Reflecting optics in the diverticular eye of a deep-sea barreleye fish (*Rhynchohyalus natalensis*). *Proc Biol Sci.* 2014;281(1782):20133223. doi:10.1098/rspb.2013.3223

Page 27: Dolphin ear
Simo H, Nummela S, Reuter T. Anatomy and physics of the exceptional sensitivity of dolphin hearing (Odontoceti: Cetacea). *J Comp Physiol A Neuroethol Sens Neural Behav Physiol.* 2010;196(3):165-179. doi:10.1007/s00359-010-0504-x

Plencner T. *Is the hearing of whales and dolphins fully developed at birth? An investigation of the odontocete inner ear.* Master's thesis. University of Otago. 2018. Accessed September 1, 2020. https://ourarchive.otago.ac.nz/handle/10523/7992

Page 28: Sagittal section of spine
This illustration was in part adapted from:
Vertebrae–normal [illustration]. *Atlas of Veterinary Clinical Anatomy.* Hill's Pet Nutrition. Accessed September 15, 2020. https://www.hillsvet.ca/en-ca/practice-management/atlas/vertebrae#VertebraeNormal

Page 29: Dorsal view of dog cervical vertebrae
Fernandes R, Fitzpatrick N, Rusbridge C, Rose J, Driver CJ. Cervical vertebral malformations in 9 dogs: radiological findings, treatment options and outcomes. *Ir Vet J.* 2019;72:2. Accessed September 16, 2020. doi:10.1186/s13620-019-0141-9

Widmer WR, Thrall DE. Canine and feline vertebrae. In: Thrall DE, ed. *Textbook of Veterinary Diagnostic Radiology.* 7th ed. Saunders; 2018:249-270.

Page 30: Giraffe cervical vertebrae
Danowitz M, Solounias N. The cervical osteology of *Okapia johnstoni* and *Giraffa camelopardalis*. *PLoS One.* 2015;10(8):e0136552. doi:10.1371/journal.pone.0136552

Badlangana NL, Adams JW, Manger P. The giraffe (*Giraffa camelopardalis*) cervical vertebral column: a heuristic example in understanding evolutionary processes? *Zool J Linn Soc.* 2009;155:736-757. doi:10.1111/j.1096-3642.2008.00458.x

Page 31: Turtle cervical vertebrae
These illustrations were created using reference CT scans from:
Werneburg I, Hinz JK, Gumpenberger M, Volpato V, Natchev N, Joyce WG. Modeling neck mobility in fossil turtles. *J Exp Zool B Mol Dev Evol.* 2015;324(3):230-243. doi:10.1002/jez.b.22557

Page 31: Snake
The body of the snake was adapted from a public domain image:
Lydekker RA. Skeleton of snake [illustration]. The Royal Natural History. Vol. 5; 1896. Thomas Fisher Rare Book Library, University of Toronto. Accessed October 14, 2020. https://commons.wikimedia.org/wiki/File:SnakeSkelLyd.jpg

Page 32: C13 and T1 vertebrae
These illustrations were created using reference micro-CT scans from:
Krings M, Nyakatura JA, Fischer MS, Wagner H. The cervical spine of the American barn owl (*Tyto furcata pratincola*): 1. Anatomy of the vertebrae and regionalization in their S-shaped arrangement. *PLoS One.* 2014;9(3):e91653. doi:10.1371/journal.pone.0091653

Pages 36, 37, 39: Reptile heart anatomy
Wyneken J. Normal reptile heart morphology and function. *Vet Clin North Am Exot Anim Pract.* 2009;12(1):51-63, vi. doi:10.1016/j.cvex.2008.08.001

Page 37: Shark circulatory system
De Iuliis G, Pulerà D. The shark. In: De Iuliis G, Pulerà D, eds. *The Dissection of Vertebrates.* 2nd ed. Academic Press; 2011:27-77.

Page 38: Octopus circulatory system
Johansen K, Martin AW. Circulation in the cephalopod, *Octopus dofleini*. *Comp Biochem Physiol.* 1962;5:161-176. doi:10.1016/0010-406x(62)90102-0

Lockey R. Octopus hearts [digital illustration]. *BBC Sci Focus Mag.* Accessed October 21, 2020. https://www.sciencefocus.com/nature/why-does-an-octopus-have-more-than-one-heart

King AJ, Henderson SM, Schmidt MH, Cole AG, Adamo SA. Using ultrasound to understand vascular and mantle contributions to venous return in the cephalopod *Sepia officinalis* L. *J Exp Biol.* 2005;208:2071-2082. doi:10.1242/jeb.01575

Page 38: Cockroach circulatory system
Fig. 18.57C: Blood vascular system of *Periplaneta* [illustration]. Essay on the circulatory system of cockroach. Biology Discussion. Accessed October 28, 2020. https://www.biologydiscussion.com/essay/essay-on-cockroach/34180

Specific Reference Photo Credits:

Cover: Gold dust day gecko (*Phelsuma laticauda*)
Credit: Tory Kallman / Flickr

Page 4: Indian elephant skull (*Elephas maximus*)
Credit: Dave King / Dorling Kindersley / Science Source

Page 7: Glyptodon skull
Credit: Kevin Walsh / Flickr

Page 10: Hoatzin (*Opisthocomus hoazin*) chick
Credit: Flip de Nooyer / Minden Pictures / SuperStock

Page 23: Brownsnout spookfish
Credit: Danté Fenolio / Science Source

Page 30: Three-toed sloth (*Bradypus variegatus*)
Credit: Juan Carlos Vindas / Getty Images

Page 33: Seven intact dolphin neck vertebrae
Credit: Elizabeth Leyden / Alamy Stock Photo

Page 33: West Indian manatee mother and calf skeletons (*Trichechus manatus*)
Credit: Millard H. Sharp / Science Source